signed

for Laryn

"Mr. Sam" ponders a point while editing a manuscript. Samworth was in his late 50s when this self-portrait was taken. Photo courtesy Mr. Abbie V. Johnson.

ERRATA

p. 54 line 8, "nine" to read "*ten*"

p. 174 2nd paragraph, line 5, to read "originally sold *with* two different jackets; ..."

p. 274 The two entries under SUBSEQUENT PRINTINGS AND ISSUES are transposed.
Also, the first entry is a later impression and should read:
"c1945 Thomas G. Samworth Georgetown, SC. *1955* ad page date."
The second entry correctly describes the second issue of the first impression.

p. 284 lines 2, 3, to read "... Finishing *of* Gunstocks, and Checkering and Carving *of* Gunstocks, ..."

p. 306 3rd paragraph heading to read "*Whelen*" (with apologies to the late Colonel)

Errors do not "creep into" a work; they simply are not removed. The compiler is responsible and regrets the need for this slip.

SAMWORTH BOOKS:

A DESCRIPTIVE BIBLIOGRAPHY

Thomas G. Samworth and the Small Arms Technical Publishing Co.

by

BRIAN R. SMITH

Foreword by Dr. Jim Casada

A guide for the use of librarians, researchers, collectors, used and rare book dealers and others with an interest in American firearms technical literature.

Brian R Smith

20/100

1 September 1990

The Marksman's Bookshelf
P.O. BOX 93
LUKE, MD 21540

SAMWORTH BOOKS:

A DESCRIPTIVE BIBLIOGRAPHY

Thomas G. Samworth and the Small Arms Technical Publishing Co.

Printed and bound in the United States of America.

Published by:

The Marksman's Bookshelf
P.O. Box 93, Luke, MD 21540

Library of Congress Catalog Card No. 90-61552

ISBN 0-9626820-8-X cloth hardbound edition
ISBN 0-9626820-9-8 leather bound limited edition

FIRST EDITION

First Printing limited to 1000 copies
900 cloth hardbound
100 leather bound limited edition,
signed and numbered

TABLE OF CONTENTS

ACKNOWLEDGEMENTS

This reference would not exist but for the generous assistance of several individuals who share a passion for books in general and Small Arms Technical Publishing Co. titles in particular. For once, space does permit mention of those to whom the compiler is indebted:

Certainly the greatest note of thanks goes to my wife Carol for her support, criticism, and patience; and for retyping the manuscript into a PC compatible word processor.

Gratitude is expressed to Mr. H. Paul Dove, Director of the James A. Rogers Library at Francis Marion College in Florence, SC, for use of biographical information previously published, and for the opportunity to photograph dust jackets on books held in the Rogers Library Special Collections. Mr. Dove also contributed additional anecdotes about Thomas G. Samworth, and provided many helpful criticisms and suggestions.

The compiler is indebted to Mr. Abbie V. Johnson of Georgetown, SC for his generosity in providing the photographs of Samworth, plus much background and reference material on Mr. Sam and his business. Mr. Johnson was a close personal friend of Mr. Sam, and was executor of his estate.

Dr. Jim Casada of Winthrop College in Rock Hill, SC, a renowned sporting book and Africana authority, graciously offered to proof the manuscript, after which he penned the very kind words found in the Foreword. Dr. Casada is the author of numerous articles and sporting books, and has published biographical pieces on Samworth that have appeared in *South Carolina Wildlife* magazine and *Shooter's Bible.*

Mr. F. Phillips Williamson of Cambridge, MD and San Antonio, TX provided additional background and personal anecdotal information which greatly enrich this work. Mr. Williamson was closely associated with Mr. Sam from WW II on. He is a recognized authority on sporting books in general, having several reference articles to his credit, and owns perhaps the finest privately-held collection of Samworth books and related material.

Jeff Brown of Aspen Hills Books was very kind to proof the manuscript and supply numerous suggestions, criticisms, as well as being an excellent "sounding board" for ideas.

Lee Jarrett, a sporting book dealer of Maysville, GA deserves credit for proofing the manuscript and calling to attention many features about specific titles. His sharp eye for detail, as well as comments and suggestions, enhance the content of this work.

The compiler wishes to thank Mr. and Mrs. Richard J. Smith, his parents, for using their mastery of the English language in proofing the manuscript. They are responsible for the existence of this work, years ago having instilled the compiler with a love for the printed word.

M. L. Biscotti, who also has one of the finest private collections of Samworth books known to the compiler, graciously provided an opportunity to photograph several mint condition dust jackets, and examine his impressive collection for variations.

Larry Williams of Gahanna, OH supplied Samworth sales literature in addition to several suggestions and constructive criticisms.

Lilo Eder of Ft. Ashby Books in Ft. Ashby, WV was extremely helpful by supplying the compiler with valuable material on the general book trade, and by special-ordering current reference books.

Other sporting book dealers and collectors who have directly or indirectly contributed to this work include: Fred M. Talkington; Gary Braithwaite; Judith Bowman; Frank Mercatante; Tom Rowe; Joe Riling; Dave Wheeler; Bob Borcherdt; Phil White; Melvin Marcher; Phil Ames; David Schwartz of Dutchman Books; Henry Paul of Game Bag Books; Michael Schnitter of Q. M. Dabney, Inc; Martin Beck; Don Mason; John Valle; Norm Berringer; and J. M. Allen.

Finally, Jerry Williams of Precision Printing, Inc. of Cumberland, MD was at once extremely helpful to and patient with a novice publisher; the quality of the physical volume you now hold is the product of his talents and those of his capable staff.

FOREWORD

Thomas Garvine Samworth, universally known in his wide circle of acquaintances as "Mr. Sam," was an individual who truly deserved the often overworked description "character." A crusty curmudgeon whose mastery of foul language would have shamed the saltiest sailor, "Mr. Sam" was also arguably the most knowledgeable authority on guns which twentieth-century America has produced. He had the distinction, now largely forgotten, of revamping the National Rifle Association's house magazine and giving the work its current title, *The American Rifleman*, and his was an imaginative, innovative mind which made important contributions while he worked for Du Pont in the munitions industry. Furthermore, beneath his gruff and rather imposing exterior lay a kind, gentle inner man with a deep respect for wildlife and the American sporting tradition.

Indeed, in many ways his love of nature and sport were his life, and it was only appropriate that he earned his livelihood from a publishing business closely associated with his loves. His Small Arms Technical Publishing Company was a one-man enterprise, but in the course of almost three decades of existence, Mr. Sam's Company produced forty-seven volumes which have come, with the passage of time, to be recognized as singular contributions to the corpus of sporting firearms literature.

Some of the books are true classics. For example, Henry Edwards Davis' *The American Wild Turkey* is without question the finest book ever written on the majestic baron of the hardwood sloughs, piney ridges, and backwoods pasture edges. Similarly, John "Pondoro" Taylor's *African Rifles and Cartridges* is a work which must be a foundation stone in any representative collection of sporting Africana. Other Samworth imprints, such as Kenneth Fuller Lee's *Big Game Hunting and Marksmanship* and Elmer Keith's *Big Game Rifles and Cartridges* are also rightly renowned.

Yet for all the undeniable importance of these books, they are but a sampling of a list which was of uniformly high standard and enduring value. Anyone who delves deeply into Small Arms imprints soon comes to realize that Samworth was not only a genius at coming to contractual terms with accomplished, authoritative contributors; he was also an exacting, excellent editor.

In recent years insightful collectors have come to realize as much, and as a result Small Arms publications have skyrocketed in price even as they make increasingly infrequent appearances in the listings of out-of-print booksellers. Yet it is not too late for determined bibliophiles, even those of limited means, to form a representative collection of Samworth books. Now is the time to act, for I confidently predict that in time to come volumes bearing the Small Arms Technical Publishing Company imprint will be sought after with much of the same avidity now directed to works printed by Eugene Connett's famed Derrydale Press or under the Roland Ward imprint in England.

Until the appearance of the current work, however, Small Arms books were a veritable bibliographic quagmire. There were so many editions, impressions, issues, and variants that it required considerable expertise to realize exactly what one possessed or sought to acquire. Now, thanks to the singular efforts of a long-time Samworth fan, Brian R. Smith, this situation has changed. In this work we have a reference tool which is essential for anyone who is or desires to become a serious Small Arms collector.

Logically organized, lovingly prepared, and meticulously researched, *Samworth Books* fills what heretofore had been a gaping hole in reference material on firearms-related sporting books. It offers not only detailed information on the books published by Samworth; but its contents also provide us considerable biographical insight on the somewhat mysterious figure behind the books. Likewise, the authors in the Samworth stable are covered in considerable detail, and we have physical descriptions of the books and halftones of the dust jackets along with the totality of current bibliographic knowledge on Small Arms publications.

As its compiler readily admits, this work is not complete. There are simply too many unanswered questions, too many missing records, for such to be the case. Rest assured though, that this descriptive bibliography is a remarkably comprehensive one, and it will serve as a highly functional compass guiding users through the mystifying maze of Samworth imprints.

Thanks to this work, we now can join that grand old man, Mr. Sam, in the books he so lovingly crafted and left to sporting posterity. We also know more of the man and his milieu, and with each new tidbit of information he becomes more intriguing. It is likely that the entire Samworth saga will never be known or written, but with the appearance of *Samworth Books: A Descriptive Bibliography,* one undeniable fact becomes clear. Here is a volume in which we can vicariously join the sage of Dirleton Plantation along with gaining appreciable insight into his publishing business.

We stand in Brian Smith's debt, just as we owe Mr. Sam a great deal for his efforts as a conservationalist, nature lover, and publisher. This work epitomizes what a reference book of its genre should be, and in time it will become as treasured, as essential to Samworth enthusiasts, as the Small Arms Technical Publications themselves. One suspects that Mr. Sam, in whatever sporting realms or Valhalla he now moves, would look upon the work with bemusement (he never understood why others were so interested in him) along with a quiet sense of pride in the fact that his publishing efforts have been so nobly recognized.

Jim Casada
Rock Hill, SC
April 24, 1990

S.A.T.P.CO.

INTRODUCTION

The information in this book was compiled for the general use of book collectors, librarians, rare and antiquarian book dealers, and researchers, and it is specifically intended to be used as a guide in the acquisition of a representative collection of Small Arms Technical Publishing Co. (SATPCO) titles. To anyone engaged in the pursuit of book collecting, a reference is absolutely necessary in order to acquire the best possible specimens within means.

Because of the dearth of information heretofore on titles bearing the SATPCO imprint, they are not widely recognized outside the firearms, hunting, and military used book trade.

The compiler has nearly fifteen years of personal experience with SATPCO books in the building of his collection as well as those of others. Every opportunity was taken to examine any and all copies of titles discovered in used and antiquarian shops, others' collections, antique stores, libraries, garage sales, book fairs, gun shows, dealer catalogs, thrift shops and the like.

The quest began in 1973 with the purchase of a 1935 work titled *A Rifleman Went to War* by Capt. H. W. McBride, the compiler then having an interest in narratives of soldiers in combat. After reading the book several times, attention was paid to the list of titles advertised on one of the rear flyleaves. It was believed that some of these others would be equally interesting, and from that tiny flake a massive snowball formed.

Along the way a few mistakes were made; some very rare titles were inadvertently passed up, other relatively common titles unknowingly carried irrationally high price tags, yet were purchased anyway. More disturbing, a few titles were acquired in the mistaken belief that they were "first impressions" or "first editions", being so represented when in fact it turned out they were not. So others may benefit from his mistakes, the compiler thus offers the product of his experience and observations.

The difficulties facing the present-day enthusiast arise from the fact that the Small Arms Technical Publishing Co. was largely a one man show, with owner-publisher Thomas G. Samworth performing all duties from soliciting and editing manuscripts to packing shipments of books to dealers and customers. "Mr. Sam's" records have long been distributed to heirs and friends, and some are now perhaps destroyed; this makes it almost impossible to determine the length of print runs with respect to some titles. The compiler has examined dozens of Samworth's letters, purchase orders, sales slips, etc. in the pursuit of researching this reference. The material that could be obtained and compiled illuminates the largest portion of Samworth's operation.

It can be argued at this point that the business of identification of an issue of an impression of an edition is unimportant concerning SATPCO titles due to their limited and specialized topics, and the comparatively small number of copies in any impression relative to the thousands of copies for a book being published for general distribution. This is a point well taken, but bibliophiles will be book lovers and if a first issue of a first impression of a first edition exists, someone will want to collect it as such.

In the absence of complete tangible records, for these questions to be answered one must turn to available concrete and circumstantial evidence and make educated deductions based on existing information.

The bibliography is presented in the following format: A reference number is assigned to each title, not differentiating between editions, impressions and issues of a given title. Titles are listed alphabetically by author, and in the instance of multiple titles by a given author, the titles are placed chronologically by copyright date.

The entry for each title provides the best available data regarding copyright date and location, physical size, description of binding, table of contents, contemporary promotional advertising copy, a list of known subsequent impressions and/or issues, a list of subsequent editions if any, and finishes with the compiler's personal notes for that title.

Two alphabetical lists, one by author and the other by title, immediately follow the introduction. For each author or title the respective reference number and page number are given where the bibliographic information can be found.

A third list shows the titles in chronological order by publication date. The chronology was developed from data gleaned from Small Arms Technical Publishing Co. records, as well as the files of the U.S. Library of Congress Copyright Office. These lists are convenient "cross references" for quickly locating a particular title or author.

This bibliography distinguishes between editions and impressions, and both terms should be defined. An *edition* is generally accepted to be a number of copies published at a given time at a particular location and having a particular editorial content. A given edition can have a number of *impressions*, or printings.

The total number of copies in an impression may not be released at once, giving rise to more than one *issue*. The physical characteristics used to determine a first issue of an impression from later ones are known as *points* of an issue, or simply "points."

The compiler has taken liberty only with the definition for edition. The Small Arms Technical Publishing Company operated from four locations during its lifetime, and numerous titles were published in more than one of the locations. In some cases, the different locations were explicitly given on the title pages of early and late examples of a given title, but in many instances they were not. Therefore, the compiler defines an *edition* as a title published at any of the four locations by SATPCO with no significant changes in editorial content. Subsequent editions are defined as being stated as such in the books themselves, or as a given title that has been subsequently reprinted by another publisher.

Subsequent impressions are readily identified by physical appearance, publication location, or other points. It is much more difficult to delineate the number of issues in a given impression, but where this has been determined beyond reasonable doubt, as in the use of limited SATPCO records, or obvious differences in paper stock and binding materials, the compiler has done so.

It is hoped that the user understands and appreciates the difficult nature of determining complete information on all titles. This work is not represented as "complete" as there are doubtless other impressions, issues, binding variants, etc. of which the compiler is not aware, and he invites the reader with additional information to write the publisher.

Thomas G. Samworth was an innovator who made his mark on American firearms literature. He single-handedly created the firearms technical genre, serving the market that would be later served by such imprints as Stackpole and Heck (later Stackpole Publishing Co.) and their Military Service Publishing Co. division, Standard Publications, and A. S. Barnes to name a few. Perhaps *Samworth Books:* will generate the recognition for Mr. Sam that he deserves.

Information for this reference was solicited from many sources, for collection and editing by the compiler. Being human, it is inevitable that mistakes have been made. Therefore, any errors or omissions are his sole responsibility.

AUTHOR INDEX

TITLE INDEX

Titles are followed by respective reference numbers and page numbers where their descriptions can be found.

PUBLICATION CHRONOLOGY

1926 *Handloading Ammunition*
1927 *Wilderness Hunting and Wildcraft*
1927 *Pistols and Revolvers and Their Use*
1927 *Small Bore Rifle Shooting*
1928 *Modern Gunsmithing*
1929 *Modern Shotguns and Loads*
1932 *The Book of the Springfield*
1932 *Military and Sporting Rifle Shooting*
1932 *.22 Caliber Rifle Shooting*
1933 *Modern Gunsmithing 2nd edition*
1935 *Textbook of Pistols and Revolvers*
1935 *Textbook of Firearms Investigation, Identification and Evidence*
1935 *A Rifleman Went to War*
1936 *The Woodchuck Hunter*
1936 *Telescopic Rifle Sights*
1936 *Big Game Rifles and Cartridges*
1936 *Sixguns and Bullseyes*
1936 *Sixgun Cartridges and Loads*
1936 *Firearm Blueing and Browning*
1937 *Handloader's Manual*
1937 *Automatic Pistol Marksmanship*
1938 *Elementary Gunsmithing*
1938 *English Pistols and Revolvers*
1940 *Advanced Gunsmithing*
1941 *Big Game Hunting and Marksmanship*
1941 *The Inletting of Gunstock Blanks*
1941 *The Shaping of Inletted Blanks*
1941 *The Finishing of Gunstocks*
1943 *Textbook of Automatic Pistols*
1944 *Telescopic Rifle Sights 2nd edition*
1945 *The Gunsmith's Manual*
1945 *Small Arms Design and Ballistics Vol. I*
1946 *Modern American Pistols and Revolvers*
1946 *Modern American Rifles*
1946 *Professional Gunsmithing*
1946 *Small Arms Design and Ballistics Vol. II*
1947 *English Guns and Rifles*
1947 *Twenty-Two Caliber Varmint Rifles*

1947 *Shots Fires in Anger*
1948 *Ordnance Went Up Front*
1948 *African Rifles and Cartridges*
1948 *With British Snipers to the Reich*
1949 *The American Wild Turkey*
1949 *Gunstock Finishing and Care*
1950 *Hunting With the Twenty-Two*
1950 *Gunsmithing*
1951 *The Book of the Springfield 2nd edition*
1952 *Checkering and Carving of Gunstocks*
1953 *Ballistic Records*
1954 *Principles and Practice of Loading Ammunition*

This chronology was compiled using records of the Small Arms Technical Publishing Co. and the Copyright Office of the U. S. Library of Congress.

THOMAS GARVINE SAMWORTH

It is not surprising that firearms literature was to be the destiny of the son born to Fred and Elizabeth Samworth in Wilmington, Delaware on October 25, 1888. Thomas Garvine Samworth grew up on the modest family farm in the northern Delaware settlement of Pike Creek, not far from Marshallton, and at an early age he was introduced to firearms. He quickly demonstrated proficiency with a rifle, along with a preference for any activity in which one might be required.

Opportunities presented themselves with regularity on eighty-plus rural acres, allowing young Samworth to polish his skills. In his later years, "Mr. Sam" would frequently mention that the only limiting factor in his early firearms experiences was the supply of ammunition. Since he had inherited his parents' frugality, he always chose to spend pocket change on .22s for his single-shot rifle rather than sweets or tobacco.

One of his earliest full-time jobs was that of printer's assistant and later printer with a small firm in Roycrofter, New York. About 1908, he enlisted in the Delaware National Guard, eventually attaining the rank of first lieutenant.

On July 31, 1913 he married the former Dora A. Cloud of Wilmington. Their first of two daughters was born on August 10, 1914 and named after her mother. Their second child, Mary, was born on February 17, 1917.

Samworth resigned the Delaware National Guard to enlist with the U.S. Army in 1916, just in time to be shipped to Texas. Promoted to lieutenant, Samworth was attached to the Sixth Cavalry which took part in the Punitive Expedition into Mexico.

The Fates had Samworth serve out his hitch without being sent to Europe for the First World War, which he always regretted. While in the Army, however, the young officer had ample exposure to military firearms and he melded the information so learned with that of his early experiences back home on the farm. The aggregate knowledge gave way to some ideas for improvements in munitions, and after his Southwest tour was completed, he was assigned to E. I. Du Pont de Nemours in Wilmington. Several of his ideas were incorporated into the production of smokeless powder and ammunition production and were hailed as revolutionary. After mustering out, he was offered a full-time position with the chemical giant, and he continued to make suggestions and improvements in munitions, becoming recognized as an authority on the subject.

Despite his success, he was too much his own man; too strong-willed and single-minded to be acclimated to corporate politics. The Samworth characteristic of speaking his mind did not mesh well with some Du Pont organization men, who took exception to his frank outspokenness. A decision was made to part company, and after a brief stay at Pike Creek, Samworth relocated to Washington, D.C. and the fledgling National Rifle Association.

During his tenure at Du Pont, Samworth's successes in the ammunition field caused him to be recognized nationally, manifested in one way by being elected a director of the NRA. Col. Sandy McNabb was the president of the NRA in 1921, which had a paltry 2000 members at the time. Director Samworth strongly believed in the organization and the need to recruit additional membership. The journal of the NRA was called *Arms and the Man,* and Samworth had ideas for improvements that might stir up interest. McNabb invited him to Washington to take over as editor of the publication, which Samworth promptly did.

In addition to modifying the editorial content, one of the first things Samworth did was to change the name of the magazine to *The American Rifleman.* The new title reflected the new spirit that was desired by

The American Rifleman

EDITOR
THOMAS G. SAMWORTH
Advertising Manager, J. R. MATTERN
Art Editor, C. J. SMITH

Published semi-monthly on the first and fifteenth days at 1108 Woodward Bldg., Washington, D. C., by The National Rifle Association

Editorial Banner From 1926 *American Rifleman*

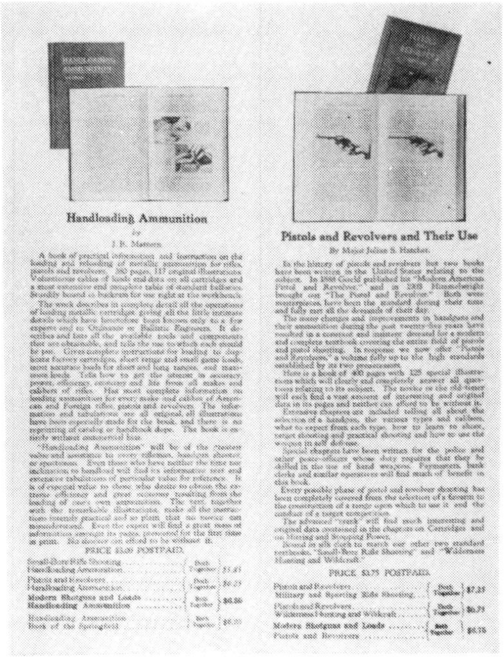
Handloading Ammunition

Pistols and Revolvers and Their Use

Two Pages of a Marshallton SATPCO Flyer

the editor. Samworth also attracted leading contemporary authorities as regular contributors. Articles appeared with bylines of such notables as Townsend Whelen, Julian Hatcher, J. Randall Mattern, Charles Askins, Sr, Ned Crossman, and others. Their knowledge and expertise was made available to the members of the young NRA, with a resultant increase in both membership and interest in the shooting sports.

Samworth thought that reasonably priced hardcover reference books might be another method of increasing interest, and with Col. McNabb's reluctant blessing, he collected six articles written by Townsend Whelen and previously published as a series in *The American Rifleman.* Titled *Amateur Gunsmithing,* the book was offered in 1925 and turned a modest profit of a few hundred dollars, with the first printing selling out quickly.

Colonel McNabb was not impressed, and forbade Samworth any further hardcover publishing ventures. Samworth forcefully restated his position on the subject, citing the financial success of the initial endeavor, and the matter of additional forthcoming manuscripts. President McNabb would have none of it, and once again Samworth steadfastly refused to yield on his principles.

He moved back to his Delaware farm and immediately initiated proceedings to have published the second manuscript he solicited while at the NRA, an authoritative work by J. Randall Mattern on reloading that was appropriately titled *Handloading Ammunition.* It appeared in December, 1926 under a new imprint; the Small Arms Technical Publishing Company of Marshallton, Delaware was born.

Samworth was assisted by Townsend Whelen in the first few years of the endeavor. The compiler has limited evidence to prove it, but believes both men had a financial interest in the business, inferring from letters between the two written in late 1926 and early 1927. Additionally, a quotation dated July 21,

1926 from George F. Lasher Printing Co. of Philadelphia on the estimated printing costs of two titles was addressed to Samworth & Whelen of Washington, DC. For 3000 each hardbound copies of "American Wilderness Hunting and Wildcraft" and "Handloading Ammunition," the price per copy was $1.00 and $.96 respectively!

The initial plan was to market the Mattern together with Whelen's *Wilderness Hunting and Wildcraft,* as it became titled, and Hatcher's *Pistols and Revolvers and Their Use,* with more to follow. They also intended to sell firearms- and hunting-related books written and published by others by advertising in magazines and using the mail-order marketing approach.

Whelen wrote to Samworth the following in late December of 1926:

"Herewith the letter and folder you asked for. I think that it looks all right, but you go over it and make any changes you think best in it and fire it in to Taylor to print.

"Better also send me at once a supply of the kind of envelopes that you will mail it in so I can begin to address these envelopes from my list. This is going to be one hell of a job, and the sooner I get started on it the better. Perhaps Taylor had better print us a new supply of envelopes for this letter and circular, and for our large catalogue. Also please consider in this connection whether we had not better delay in sending the letter and circular out for a week until we can get our big catalogue printed.

"You say that you think that we ought to list a lot more books in our catalogue. I think that you wrote this before you got my letter sending you the samples of the first pages of the catalogue, and the list of books I proposed to include. You will see that I gave long descriptions to only about one-third of the books we propose to list, and the remaining (titles) are to be placed at the end. In this list there are a large number of African books, being the very best books on the subject that are now in print. The trouble is that a number of the best books on Africa are now out of print, and this applies to books on other subjects as well . . . we can put a notice at the end to the effect that we can procure any book the customer wishes. I think that we had better let this catalogue go out to include just the books I have listed. It will take a long time to write the publishers relative the sales price and commission on other books, and this will greatly delay getting out the catalogue. We want to fire that catalogue out now while the iron is hot. I am for sailing ahead with it quick as is."

The list of books proposed by Whelen included:

Handloading Ammunition, Mattern; *American Wilderness Hunting and Wildcraft*, Whelen; *Small Bore Handbook*, Crossman; *Modern Pistols & Revolvers and Their Use*, Hatcher; *Camping and Woodcraft*, Kephart; *Wilderness of the Upper Yukon*, Sheldon; *White-Tailed Deer*, Newsom; *Game Trails in British Columbia*, Williams; *Life of F. C. Selous*, Millais; *The Arctic Prairies*, Seton; *The American Rifle*, Whelen; *Amateur Gunsmithing*, Whelen; *American Shotgun*, Askins; *Sporting Firearms Today In Use*, Curtis; *Sporting Rifles and Rifle Shooting*, Caswell; *Camp Fires in the Canadian Rockies*, Hornaday; *Camp Fires on Desert and Lava*, Hornaday; *On the Headwaters of the Peace River*, Haworth; *My Life With the Eskimos*, Stefansson; *The Friendly Arctic*, Stefansson; *African Game Trails*, Roosevelt; *The Land of the Lion*, Rainsford; *African Camp Fires*, White; *The Land of the Footprints*, White; *The Rediscovered Country*, White; *Lions in the Path*, White; *The Adventures of an Elephant Hunter*, Sutherland; *Big Game Hunting in the Himalayas & Tibet*, Burrard; *East of the Sun and West of the Moon*, K.& J. Roosevelt; *Through the Brazilian Wilderness*, Roosevelt; *Scouting on Two Continents*, Burnham.

The reader will no doubt recognize some or all of the titles; note the second through fourth books listed were SATPCO titles in production at the time and each saw a minor change in title.

They did advertise a number of the above in early SATPCO sales literature, but eventually dropped all of them when sales went flat with the Depression. One of the earliest announcements for SATPCO books took the form of an open letter of January 1, 1927 written on Company stationery:

To the Riflemen, Hunters, and Sportsmen;
United States of America

Gentlemen:

May we call your attention to the organization we have formed to cater to your needs, and to supply you with practical, interesting, helpful literature covering your favorite sport? It is our intention to publish the very best books on your subjects by the leading authors, and to offer them to you at prices far lower than such books have ever been offered before--lower even than pre-war prices.

Today we want particularly to tell you that we are just about to bring out a very remarkable new book by Lt. Colonel Townsend Whelen entitled "Wilderness Hunting and Wildcraft", an intensely practical, and interesting work that is entirely different from anything else ever written on the subject, and by far the best thing that has ever come from the pen of Colonel Whelen. It will be ready for delivery by February 15th, and a prospectus of it is enclosed.

Now here is our proposition:

By special arrangement with the author we are able to offer you this book, specially autographed by Colonel Whelen with pen, at the regular price of $3.75 if you will get your order to us before February fifteenth. This is a wonderful opportunity. Do not neglect it. Get in your order today, using the order blank on the prospectus, and the autographed copy will be mailed to you the day it comes off the press.

The proofs are now being read, and we expect to have it ready by February fifteenth at the very latest. Act now and get your autograph copy!

May we also not send you at the same time the new work by Mr. J. R. Mattern; "Handloading Ammunition." This book is proving a very great success. We have had hundreds of letters from riflemen attesting to its worth and helpfulness, and not one single complaint. It gives absolutely all the dope on the subject, covering every rifle, every cartridge, every bullet, every load. It is simply fine, as all our books will be.

We are also bringing out other works of exceptional value, books that you have always wanted. Watch our advertisements, circulars, and announcements. It will pay you if you want to be abreast of the times, get the most from your favorite sport, and save money.

Fraternally yours,
Small Arms Technical Publishing Co.
by T. G. Samworth

It was a risk of no small proportion, but there was a steadily growing demand for sporting books in the United States at the time. Most quality literature on hunting and shooting was published in Europe and Great Britain in those years, a fact that escaped neither Mr. Sam, nor another famous sporting book publisher, Eugene V. Connett of Derrydale Press fame.

While Connett steered his Derrydale operation toward sporting books centering on hunting, horses, and dogs, Samworth chose to concentrate on what he knew best: the technical aspects of firearms and ammunition. The gamble paid off; the market for firearms technical books was there all along, just as Samworth had so resolutely believed.

In retrospect, it was a dormant market that he exploited; Samworth merely awakened it by making his wares available.

Handloading Ammunition was hailed as a definitive work, being the only contemporary compilation of reloading knowledge of its length and content. Samworth immediately followed with more titles, including the previously mentioned *Pistols and Revolvers and Their Use* by Julian Hatcher, joined by *Modern Gunsmithing* by Clyde Baker, Charles Askins, Sr's *Modern Shotguns and Loads*, and *Small Bore Rifle Shooting* by E.C. Crossman.

While these early Small Arms Technical Publishing titles were being introduced, Samworth based his operation at Marshallton, DE, close to Wilmington. The post-WW1 years saw an economic boom bring prosperity and expansion to the Wilmington area, and the city grew to envelop Pike Creek and Marshallton. Samworth watched with distaste as his boyhood hunting sanctuary yielded to inevitable development. Since he had a strong aversion to cold winters in general, Mr. Sam took steps to relocate in a southerly direction just as the Crash was about to hit Wall Street.

Characteristically decisive, in 1928 he used savings from his Du Pont days to purchase a 700-acre plantation in Onslow County, NC, on the New River near Jacksonville. In September, 1930 he made the agreement to sell the Pike Creek farm to one of the Du Ponts, and went to North Carolina to supervise construction of a house and office for his family and business. The sale of the farm was closed on February 15, 1931, and Samworth moved his family into their brand-new quarters immediately thereafter.

The new location was idyllic to the outdoorsman-publisher. He considered it to be primarily a shooting preserve that served a secondary function as a location for his business. Post-Depression sales were somewhat flat but steady, due to the introduction of such works as *The Book of the Springfield* and *Military and Sporting Rifle Shooting*, both by E.C. Crossman, Charles Singer Landis' *.22 Caliber Rifle Shooting*, *A Rifleman Went to War* by Capt. Herbert McBride,

S.A.T.P.CO.

Mister Reader!

This book was written by a shooter and published by shooters for shooters. These are three of the reasons why you will like it.

The Small Arms Technical Publishing Company is a rather unique organization in its way. We endeavor to deal directly with our readers and to cut out the usual middleman's profit. Therefore we were able to sell you this book at about two-thirds the price other publishers or their local agents would have charged you. Our authors, whom you may have noted are the very best in their lines, are each members of our organization as far as their book is concerned. They receive a much larger return from its sale than is customary and hence give us their very best work. And we are constantly bringing out other books of the calibre of this one, books which we know will interest you and your friends who are also shooters.

The shooting fraternity is scattered all over this country. We are selling them this book and others similar to it; real, helpful, entertaining and practical literature of the kind they have always wanted and which will give them greater assistance in their chosen sport. We want to get in touch with every one of them. Our success and the degree of helpfulness which we can be to them depends upon doing this if possible. We want to learn of their needs and be really useful to them. We will publish any book for which there is a popular demand.

Will you not show this book to your friends who are also shooters, or tell them about it, and at the same time do us the favor of handing this slip to one of them whom you think will be interested in this or other unusual shooting or outdoors books. Then he can send for it, or for our catalog, or get his name placed upon our mailing list, using form on opposite side of page for this purpose. Or better yet; send us their names.

We want to be of all the assistance possible to the riflemen, pistol and shotgun shooters, hunters and sportsmen of this country. We believe we understand just what kind of books they want and we are prepared to publish and furnish them at the very lowest prices consistent with the best authors and the highest workmanship. Lend us a helping hand! Pass along this dope will you? Let your friends see this book. Compare any of our books—the number of pages, the number of words on a page, the mass of illustrations, the substantial binding, the workmanship and praticularly the authors and the information they have given you—with other works. Then compare our prices. And remember that we pay the delivery charges.

We have other splendid works now being published or in hand for publication. Get your name and that of your friends on our mailing list, let us send you our pamphlets and also advise you promptly whenever we bring out a new book on some branch of your special hobby.

Small-Arms Technical Publishing Company

Marshallton, Delaware

Our Publications:

NOW READY:

Handloading Ammunition, *J. R. Mattern*

Wilderness Hunting and Wildcraft, *Townsend Whelen*

Small Bore Rifle Shooting, *E. C. Crossman*

Pistols and Revolvers and Their Use, *Julian S. Hatcher*

Modern Gunsmithing, *Clyde Baker*

Modern Shotguns and Loads, *Charles Askins*

IN PREPARATION:

Target and Sporting Rifle Shooting, *E. C. Crossman*

Campfire Talks, *Chauncey Thomas*

Pistol Shooting, *J. H. FitzGerald*

American Sporting Rifles, *Townsend Whelen*

Marshallton ad "slip" usually laid-in books sold
compare to Onslow slip, right

Mister Reader!

This book was written by a shooter and published by shooters for shooters. These are three of the reasons why you will like it.

The Small Arms Technical Publishing Company is a rather unique organization in its way. We endeavor to deal directly with our readers and to cut out the usual middleman's profit. Therefore we were able to sell you this book at about two-thirds the price other publishers or their local agents would have charged you. Our authors, whom you may have noted are the very best in their lines, are each members of our organization as far as their book is concerned. They receive a much larger return from its sale than is customary and hence give us their very best work. And we are constantly bringing out other books of the calibre of this one, books which we know will interest you and your friends who are also shooters.

The shooting fraternity is scattered all over this country. We are selling them this book and others similar to it; real, helpful, entertaining and practical literature of the kind they have always wanted and which will give them greater assistance in their chosen sport. We want to get in touch with every one of them. Our success and the degree of helpfulness which we can be to them depends upon doing this if possible. We want to learn of their needs and be really useful to them. We will publish any book for which there is a popular demand.

Will you not show this book to your friends who are also shooters, or tell them about it, and at the same time do us the favor of handing this slip to one of them whom you think will be interested in this or other unusual shooting or outdoors books. Then he can send for it, or for our catalog, or get his name placed upon our mailing list, using form on opposite side of page for this purpose. Or better yet; send us their names.

We want to be of all the assistance possible to the riflemen, pistol and shotgun shooters, hunters and sportsmen of this country. We believe we understand just what kind of books they want and we are prepared to publish and furnish them at the very lowest prices consistent with the best authors and the highest workmanship. Lend us a helping hand! Pass along this dope will you? Let your friends see this book. Compare any of our books—the number of pages, the number of words on a page, the mass of illustrations, the substantial binding, the workmanship and praticularly the authors and the information they have given you—with other works. Then compare our prices. And remember that we pay the delivery charges.

We have other splendid works now being published or in hand for publication. Get your name and that of your friends on our mailing list, let us send you our pamphlets and also advise you promptly whenever we bring out a new book on some branch of your special hobby.

Small-Arms Technical Publishing Company

Marines, Onslow Co., North Carolina

S.A.T.P.CO.

Our Publications:

NOW READY:

Handloading Ammunition,
J. R. Mattern

Wilderness Hunting and Wildcraft,
Townsend Whelen

Small Bore Rifle Shooting,
E. C. Crossman

Pistols and Revolvers and Their Use,
Julian S. Hatcher

Modern Gunsmithing,
Clyde Baker

Modern Shotguns and Loads,
Charles Askins

Book of the Springfield,
E. C. Crossman

Military and Sporting Rifle Shooting,
E. C. Crossman

22 Caliber Rifle Shooting,
C. S. Landis

Onslow ad "slip" usually laid-in books sold
compare to Marshallton slip, left

and the ever popular Hatcher work, *Textbook of Pistols and Revolvers.* In 1932, all Small Arms Technical Publishing Co. titles cost approximately four dollars, give or take a buck. For some people, this was more than a week's wages.

Nonplussed, Samworth realized the situation for what it was and brought out the little manuals that collectors have since dubbed "Small Sams". Appearing first in 1936, the following titles sold for half the asking price of previously published SATPCO works: *Firearm Blueing and Browning* by Angier, *Woodchuck Hunter* by Estey, Elmer Keith's *Big Game Rifles and Cartridges* and *Sixgun Cartridges and Loads, Telescopic Rifle Sights* by Whelen, and *Sixguns and Bullseyes* by Reichenbach. Other titles in the same series followed over the next few years, their success a testimony to Samworth's ability to provide the enthusiast with quality firearms technical literature written by acknowledged experts all available at an affordable price.

While the little manuals sold well, economically speaking they were a draw for the publisher, as he related to Judge Charles S. Landis in a 1947 letter:

". . . with relation to your proposition of small books, my experience has been that there is no money or profit in them. I have a line of 12 which I published 13 years back and have been selling ever since, with additions thereto. I get them out to sell for $1.50, the theory being that for every one of my $5.00 books sold we would sell at least 5 of the lower-priced books. Well, it did not work out at all; and the cheaper books sold neck and neck with the higher priced ones. I dropped the scheme and now get out no books which sell for less than $3.00. The shooting and collector market is a very limited and critical one and can be reached only through certain channels. These people want to get every damn bit of data and information possible to obtain. They want none of this 'touch lightly' attitude in their books but want the full and

50

freely and voluntarily, without fear or compulsion of her said husbandand, or any other person, and that she doth still voluntarily assent thereto.
Witness my hand and seal, this the 21 day of November, 1929.

(Notarial Seal) R. W. Craft Notary Public
My commission expires the 6 day of Dec. 1930.

NORTH CAROLINA, ONSLOW COUNTY.

I, R. W. Craft a Notary Public in and for the State and [illegible] County, [illegible] hereby certify that [illegible] personally appeared before me this day and acknowledged the due execution of the annexed instrument, and the said [illegible] being by me privately examined, separate and apart from her said husband [illegible] freely and voluntarily, without fear or compulsion of her said husband, or any other person, and that she doth still voluntarily assent thereto.
Witness my hand and seal, this the 21 day of November, 1929.

(Notarial Seal) R. W. Craft Notary Public
My commission expires the 6 day of Dec. 1930.

NORTH CAROLINA, ONSLOW COUNTY.

The foregoing certificate of R. W. Craft a Notary Public of Duplin County is adjudged to be correct and sufficient. Let the instrument together withcertificate be registered.
Witness my hand and seal this the 1st day of Mar. 1930.

J. R. Gurganus, Clerk Superior Court.

Filed for registration at 10:30 o'clock A. M. March 1st, 1930 and recorded March 6, 1930.

J. E. Sanders Register of Deeds.

::

NORTH CAROLINA:
ONSLOW COUNTY:

THIS DEED, Made this 10th day of February, 1930 by Ollie Marine and Mary Marine, his wife and Kittie Marine (widow) of Onslow County and State of North Carolina, of the first part, to Thomas G. Samworth of Onslow County and State of North Carolina, of the second part:

WITNESSETH, That said Ollie Marine and wife, Mary Marine and Kittie Marine, in consideration of Five Thousand Five Hundred ($5500.00) Dollars, to them paid by Thomas G. Samworth, the receipt of which is hereby acknowledged, have bargained and sold, and by these presents do grant, bargain, sell and convey to said Thomas G. Samworth, his heirs and assigns, a certain tract or parcel of land in Onslow County, State of North Carolina and described as follows:-

Being known and designated as Wilkin's Bluff and being situate below the village of Marine and near the mouth of New River and on the Eastward side of said river;

BEGINNING at a stake about one hundred feet Westward from the old landing in the cove to the Northward of said Wilkin's Point, said stake being the dividing line between the lands formerly owned by Ollie Marine and F. L. Marshall and said stake standing on the Eastward shore of New River, and running thence North 23 degrees 57 minutes East 196-77/100 feet; thence North 23 degrees 14 minutes East 159 feet; thence North 25 degrees 13 minutes East 216-8/10 feet; thence North 36 degrees 51 minutes East 73-7/10 feet; thence North 47 degrees 46 minutes East 145-55/100 feet; thence North 43 degrees 01 minutes East 104-7/10 feet; thence North 41 degrees 45 minutes East 174-25/100 feet; thence North 44 degrees 08 minutes East 90-65/100 feet; thence North 46 degrees 30 minutes East 223-3/10 feet; thence North 48 degrees 18 minutes East 189-75/100 feet; thence North 50 degrees 23 minutes East 121-5/10 feet; thence North 54 degrees 05 minutes East 62-15/100 feet; thence North 67 degrees 01 minutes East 225-9/10 feet; thence North 81 degrees 50 minutes East 295-2/10 feet; thence South 55 degrees 49 minutes East 88-9/10 feet; thence South 47 degrees 19 minutes East 133-3/10 feet; thence South 81 degrees 56 minutes East 134-95/100 feet; thence South 83 degree 05 minutes East 298 feet; thence South 82 degrees 03 minutes East 479-6/10 feet; thence South 81 degrees 08 minutes East 271-8/10 feet; thence South 11 degrees 03 minutes West 500 feet; thence South 14 degrees 18 minutes West 252 25/100 feet; thence South 8 degrees 21 minutes West 421-4/10 feet; thence South 8 degrees 39 minutes West 228-75/100 feet; thence South 40 degrees 02 minutes West 765-2/10 feet; thence South 47 degrees 57 minutes West 127 feet; thence South 46 degrees 28 minutes West 181 feet; thence South 39 degrees 25 minutes West 188-35/100 feet; thence South 35 degrees 36 minutes West 240-4/10 feet; thence South 35 degrees 30 minutes West 162-2/10 feet; thence South 41 degrees 21 minutes West 159-2/10 feet; thence South 63 degrees 29 minutes West 103-7/10 feet; thence South 50 degrees 21 minutes West 306-1/10 feet; thence South 30 degrees 5 minutes West 94-7/10 feet; thence South 28 degrees 58 minutes West 174-5/10 feet; thence South 19 degrees 34 minutes West 273-2/10 feet; thence South 10 degrees 13 minutes West 80-2/10 feet; thence South 9 degrees 32 minutes West 72-5/10 feet; thence South 0 degrees 06 minutes West 134-6/10 feet; thence South 8 degrees 43 minutes West 119-65/800 feet to the Eastward shore of New River; thence up and with the River, its various courses, around the Westward end of Wilkins' Bluff to the beginning, containing 178-44/100 acres and being the same lands shown on a plat made by W. B. Trogdon Jr., Engineer, dated January 24th, 1930; said map having been prepared for Thomas G. Samworth.

TO HAVE AND TO HOLD the aforesaid tract or parcel of land, and all privileges and appurtenances thereto belonging, to the said Thomas G. Samworth, his heirs and assigns, to his and their only use and behoof forever.

And the said Ollie Marine and Mary Marine, his wife, and Kittie Marine, for themselves,

Deed from Ollie, Mary, and Kitty Marine to Thomas G. Samworth for 700 acres on the New River near Jacksonville, Onslow County, N.C. in the area known as "Marines." It is dated February 10, 1930.

complete material and as extensive and lengthy as possible. Each wants to be an authority on the subject of the book; hence he wants that book to contain PLENTY of material for him to absorb."

The Onslow County years saw the Small Arms Technical Publishing Co. firmly established and well recognized by the shooting community. Storm clouds were brewing, though, in the unlikely form of his "next-door neighbor."

When Samworth moved on his plantation from Delaware, he gave the address on the early Onslow books as Marines, Onslow County, North Carolina, because the property was purchased from landowners Ollie and Mary Marine. Coincidentally, another sort of "marines" were inhabiting the same locale, these being collectively the United States Marine Corps who operated the training camp that eventually became the sprawling Camp Lejeune.

To the USMC, the publisher's 700-some acres looked like prime training ground for new recruits and proceedings were started to have Samworth's ground condemned under eminent domain.

As one would expect, Mr. Sam battled the (khaki and blue variety) Marines' move tooth and nail for several years, finally ending up in court. The result was painful to pride and pocketbook; he was compensated only a third of the purchase price he had paid. Moreover, the presiding judge chastised him for contesting the government's claim to his property as being unpatriotic!

Once again demonstrating himself to be more than a match for such a devastating setback, the undaunted Samworth picked himself up by his bootstraps and in 1940 moved to South Carolina, purchasing Exchange Plantation on the Pee Dee River north of Georgetown. Samworth described his exodus from the Onslow County debacle to South Carolina as "coming to America". He was captivated by the coast-

This is one of the books published by the

SMALL-ARMS TECHNICAL PUBLISHING COMPANY

AN organization composed of sportsmen and shooters, who publish the very best works possible on shooting and kindred subjects and who constantly deal more or less directly with the marksmen of the country, be they sportsmen, big-game hunters, riflemen or explorers.

On any one subject the very best author in the world is obtained; he is made, per se, a part of the organization, and he is given every assistance as will result in the very best book on that subject possible to collate and write. The books are so prepared that they become immediately the authority on their subject the world over. And when compared with other works as to size, number of words in text and number of illustrations, they will be found to be the equal of similar books selling at twice their price.

The Company is constantly bringing out new books and it is suggested that those individuals interested in shooting and kindred outdoor books keep in touch with their local bookseller or get their name on the mailing list of:

THE SMALL-ARMS TECHNICAL PUBLISHING COMPANY

MARSHALLTON, DELAWARE

Early SATPCO promotional copy.

Edited and Published by

THOMAS G. SAMWORTH

Georgetown, South Carolina, U. S. A.

"Case Head" logo and address found on Georgetown titles. Variants of this logo were used from 1938 on.

al geography and its possibilities for the role of the quintessential "gentleman plantation owner."

Within a year an Austrian paperhanger with political ambitions and the premier of Japan teamed up to cause Samworth further difficulty. Rationing was the order of the day for stateside Americans during WW II, and paper was classified as a strategic material. Making use of his widespread connections, Samworth was able to convince authorities that his firearms technical literature would make a contribution to the war effort, and the presses went full speed ahead.

In the middle of all the editing, proofreading, packaging, and shipping, Mr. Sam once again pulled up stakes and in 1943 purchased the 600-acre Dirleton Plantation outside of Georgetown, although the mailing address remained Plantersville until 1948.

During the war and shortly thereafter several new titles appeared bearing the Plantersville SATPCO imprint. Among them were *Small Arms Design and Ballistics* by Townsend Whelen, *Shots Fired in Anger* by John B. George, and Roy Dunlap's *Ordnance Went Up Front*.

Using his plantation house as office, warehouse, and shipping department allowed Samworth to keep his overhead very low, and these savings were in turn passed on to the customer. Mr. Sam also capitalized on the free advertising made available by the otherwise blank flyleaves at the end of the text in his books. Starting with the first Onslow County books, a one-page ad appeared in the back calling attention to the availability of additional titles of interest to the student of firearms. As time passed and the number of titles in print grew, this one-page list was joined by other adjoining pages that carried detailed descriptions of particular titles.

Similar descriptions were employed on the dust jacket flaps and back panels; these were seen on nearly all SATPCO books with the exception of the very first, *Handloading Ammunition.* Samworth saw that he got the most advertising bang for the buck, even covering the inside of the dust jacket with ads in some cases, notably those with a Plantersville imprint.

At Dirleton, Samworth had another "next-door neighbor" that proved to be more sociable than the Marines had a decade earlier. Alex Quattlebaum was the owner of a construction company and Arundel Plantation, bordering on Dirleton. Samworth and Quattlebaum soon became fast friends, sharing interests and pastimes.

Quattlebaum was responsible for introducing Samworth to Henry E. Davis, the Florence, SC lawyer and author of *The American Wild Turkey.* Samworth had originally arranged for a prominent South Carolina author to write a wild turkey book, but was extremely disappointed at the result. Long a master of scatology, he left no expletives deleted in describing his displeasure with the manuscript to neighbor Quattlebaum. In the end, Samworth made Davis the same offer of $1000 for a manuscript, but when the finished product of Davis' typewriter arrived, Samworth was so delighted with it he doubled the payment.

The year 1948 saw the appearance in Samworth books of Georgetown as the location of publication after that Post Office took over service to Dirleton. It was here that Samworth produced the two books that have proven to be the most well-known and highly sought-after of all his imprints: *The American Wild Turkey* and *African Rifles and Cartridges.*

Like most people, Mr. Sam gave thought to retirement at age 65, and stopped accepting manuscripts about the mid-1950s. Earl Naramore's *Principles and Practice of Loading Ammunition* has the distinction of being the last book Samworth published. Ironically, the venture also began with a reloading handbook.

Samworth negotiated an agreement with Stackpole Publishing Company, a leading outdoors publisher,

DIRLETON PLANTATION HOUSE
photo courtesy Mr. Abbie V. Johnson

Not a peeping "Tom," but an elderly Mr. Sam enjoying a favorite and frequent pastime, observing the abundant wildlife surrounding his beloved *Dirleton.*

Samworth with two unidentified acquaintances.
photos courtesy Mr. Abbie V. Johnson

to market and distribute most of his remaining stock. Stackpole later elected to reprint several Samworth titles, a few of which are still in print and available.

While he continued for a time to edit manuscripts for Stackpole, the sellout gave Mr. Sam more time to devote to observation of the abundant wildlife surrounding Dirleton. He was an avid birdwatcher and photographer, and a champion of wildlife conservation.

He did his share of hunting, both upland birds and waterfowl as well as deer. He claimed to have killed 141 deer before hanging up his favorite whitetail rifle, a .30-'06 caliber 1903-A3 Springfield. It was this rifle that he employed to take the deer from the window of his third-story Dirleton study.

Having been forced out of prime hunting areas in Delaware and North Carolina, Samworth wanted to prevent the same thing from happening to others. In 1962 he deeded Dirleton to the South Carolina Wildlife Department for use as a preserve and public hunting area for migratory birds. One would not on the surface expect so crusty a character to be that altruistic; it was but a manifestation of his strong personal beliefs.

"Tell you why I did that," he told the late Bill McDonald of the Columbia, SC *State.* "I'm 84, going on 85, and I've only got two daughters, and they live up north. I figure I'm getting to the end of my road.

"I always enjoyed hunting and wildlife, and the sportsmen of this country were always good to me, and I wanted to do something for them. We've got all kinds of hunting here . . . it's the best in the state."

South Carolina paid him tribute on February 17, 1978 when the Pee Dee Game Management Area was renamed the Thomas G. Samworth Game Management Area. At the same ceremony, Mr. Sam was also presented with the Order of the Palmetto, the state's highest honor award.

That he truly did care about people was demonstrated by his readiness to help any and all individuals who asked. If approached by a total stranger with a question about a particular firearm, rest assured that man came away some time later after a detailed dissertation, knowing virtually all there is to know about the weapon.

This willingness to be of service led Samworth to loan his personal collection of Small Arms Technical Publishing books to acquaintances and passers-by who demonstrated an interest in reading them. Sadly, many of them never found their way home.

Even in his advanced years, he would routinely assist state wildlife personnel with such tasks as their examination of the stomach contents of ducks shot by hunters during the season. It was convenient for him to do so, having moved out of the plantation house and into a house trailer in the yard area. State wildlife biologists often used the plantation house as a base of operations, and Samworth continued to use a room on the third floor as a study. He had never learned to drive, and relied on neighbors or others for a lift when he made one of his infrequent trips away from Dirleton.

Samworth also had strong feelings about publicity, and went out of his way to avoid it. After the news broke of Dirleton being donated to South Carolina, he was approached by several newspapers for interviews, most of which were refused.

In a 1973 interview with the late Bill McDonald, Samworth accosted: "What kind of story do you want? I don't like publicity. I'm too old for it. I had enough of that when I got married . . . You boys can pretty much ruin a man. But I guess it'll be all right if you go ahead and write it right. If you don't it won't be much good for anything."

He could never understand why others would be interested in him or what he had to say, and would never fail to mention this to the reporter granted a rare interview. Indeed, Samworth had kept such a low profile that locally many people had never heard of of him or the Small Arms Technical Publishing Co.

Samworth had always prided himself on his excellent physical condition, which served him well in several instances. Well into his eighties, he managed single-handedly to extinguish a grease fire on the kitchen stove, limiting damage to some scorched areas on cabinets and ceiling. He chose not to repair the damage, and liked to point it out to visitors, presumably as proof of his prowess.

Likewise, his strong constitution paid off when he became trapped between floors in the plantation house elevator when the circuit breaker on the hoist motor tripped. After spending part of the night in the elevator, he simply kicked out a hole in the elevator car and climbed down to the floor below, escaping what could have been certain tragedy.

Nonetheless, he did experience some health problems, and the inevitable happened on December 28, 1981. Mr. Sam was given 93 years of life, and by all accounts he made the best of it he could. Time has shown him to be the single most important figure in twentieth-century firearms literature in this country, having created the firearms technical genre. Most of the Small Arms Technical Publishing Co. titles were recognized as standard reference works in their day, and several continue to hold that distinction. Nearly all titles are currently sought for collection as well as intrinsic content, and one sure measure of success is the prodigious number of reprint editions of SATPCO titles that have been issued over the years.

Retrospectively, Mr. Sam remarked: "I'd do the same damn thing over again. I'd be as cussedly independent as I possibly could be." In a conversation with H. Paul Dove, Director of the James A. Rogers Library at Francis Marion College, Samworth was asked what his thoughts were on the afterlife. His laconic reply: "I hope the hunting will be good there."

He was laid to rest at his beloved Dirleton. The day of the funeral long-time friend and estate executor Abbie V. Johnson made this observation:

"The day of Mr. Samworth's funeral, as we drove from the funeral home back to Dirleton, his final resting place, a large hawk came out of the woods and flew over the hearse for several miles to the point where we turned into Dirleton; whereupon he perched in a tree and watched the procession pass. I guess it was just coincidence but it touched me greatly. It seemed as if all of nature was paying tribute to this man at the time of his parting . . ."

Happy hunting, Mr. Sam!

SAMWORTH BOOKS

PHYSICAL CHARACTERISTICS OF SMALL ARMS TECHNICAL PUBLISHING CO. BOOKS

MATERIALS USED - PAPER

The texts of Samworth books are printed on a variety of papers and cased in six types of bindings. For purposes of discussion, some background information on these materials follows, beginning with paper:

Uncoated book paper usually has a dull finish, is slightly rough in texture, and has a creamy, off-white shade of color. This type of paper is normally found in present-day hardcover works of fiction for ease of reading the printed matter with a minimum of eye-strain.

Coated paper is usually characterized by a glossy finish, although it is readily available in a low-gloss "dull" grade and a true "matte" or "satin" finish as well. Coated paper yields excellent graphic reproduction on artwork and photographs. That is why today's catalogs and magazines are printed almost exclusively on this type of stock.

Years ago, glossy coated paper was called "super" finish, and used only for the pages containing illustrations called "plates", due to its higher cost. In most Samworth books this type of paper is seen on the pages with plate illustrations. It is noticeably "whiter" and much more glossy than the stock used for the text, which is uncoated.

One would assume that coated paper would cost more, and it does. That is the main reason Samworth and his contemporaries generally used it only for plates, but some SATPCO books made use of it for the text. Examples of Samworth books that used coated paper throughout are Landis' *Twenty-Two Caliber Varmint Rifles*, and *Principles and Practice of Loading Ammunition* by Naramore, plus all four titles in the *Firearm Design and Assembly* series.

Volumes such as these using coated paper are heavy compared to similarly sized texts made of uncoated "book" paper.

A standard ream of printing paper is 500 sheets of a given size. That size is different for coated and uncoated grades. For example, coated paper has a standard sheet size of 25 inches by 38 inches, derived from the area of sheet required to print an octavo format signature. Refer to "Book Formats and Sizes" below.

If those 500 sheets together weigh 60 pounds, that paper is said to have a 60 pound basis weight. If 500 sheets the same size of a different stock weigh 80 pounds, that paper has an 80 pound basis weight. "Eighty-pound" as papermakers and printers call it, is one and one-third times as heavy as 60 pound and noticeably thicker, but it is not necessarily one-third again as much. Papermakers refer to paper thickness as "caliper", which is the thickness of one sheet. In general, caliper increases with increasing basis weight, but it is not necessarily a linear relationship.

Printers refer to a similar property called "bulk", the overall thickness of a stack of sheets which includes a certain amount of air trapped between each. This more realistically represents the thickness of the finished product.

World War II saw material shortages of all types, and this certainly extended to paper. A number of war-time impressions of various Samworth titles are seen with a relatively thin uncoated paper with a decidedly grayish-blue shade, somewhat similar to our present day newsprint. The second impression of *Big Game Hunting and Marksmanship* as well as later printings of *Textbook of Firearms Investigation, Identification and Evidence*, and both Elmer Keith and William Reichenbach titles are printed on this inexpensive rationed book paper. Suprisingly, it seems to age no faster than pre- and post-war Samworths printed on conventional stock.

On a few titles Samworth chose to add embellishments such as untrimmed pages and pages with "deckle edges". Untrimmed pages are just that; the pages are trimmed only at the top and bottom, leaving the end with an "offset" appearance that naturally occurs when the folded signatures are "gathered" or assembled together in order. Samworth probably wanted to give certain titles an "aged" flavor, and thus made use of untrimmed edges on *Modern American Rifles*, *English Pistols and Revolvers*, and *Textbook of Automatic Pistols* to name a few.

Deckle edge refers to the "ragged" edge that paper is made with as it is formed on the paper machine. Deckle edges are nowadays trimmed off, but books from the eighteenth and nineteenth centuries are frequently seen with them because each sheet of book paper was often made individually by hand rather than continously on a machine as it is now. One example of a Samworth title with deckle edge paper is the first impression of *Modern American Pistols and Revolvers,* in an attempt to make this reprint edition faithful to the original.

MATERIALS USED - BINDINGS

Samworth used many different types of binding cloth on his merchandise: *Buckram* refers to a coarse-weave linen cloth, which may or may not be filled with a colored pigment to varying degrees. Several grades exist, depending on the thread diameter and the number of threads per inch.

Silk is the term used by Samworth to describe the cloth used on most of his early titles. It is heavily filled with pigment and embossed with a stamp that imparts very fine, parallel vertical lines resembling phonograph record grooves. The carrier material was a fine-weave cloth that is not necessarily silk; the term refers to the appearance of the finished product.

Later issues of titles originally bound in "silk" are found with *woven silk*, a term coined by the compiler

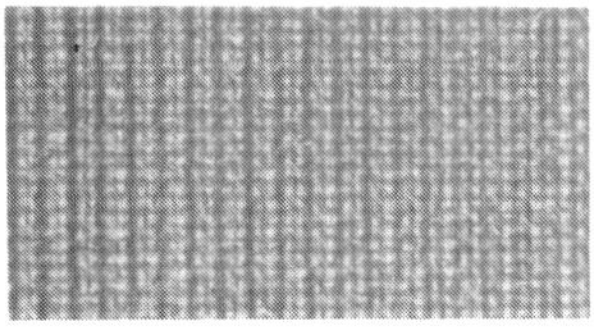

SILK

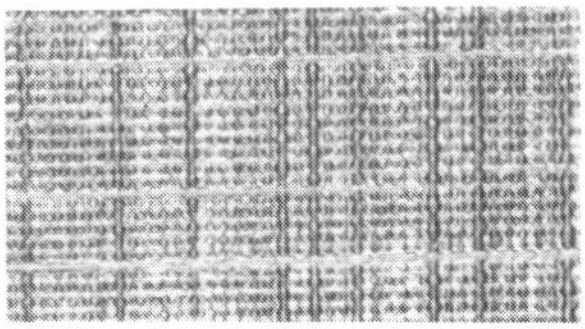

WOVEN SILK

SMOOTH CLOTH

FINE CLOTH

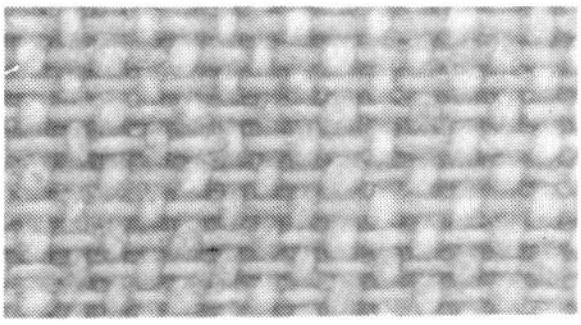

BUCKRAM

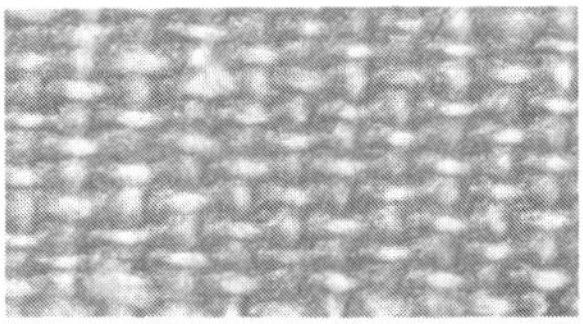

FILLED BUCKRAM

PEBBLE GRAIN

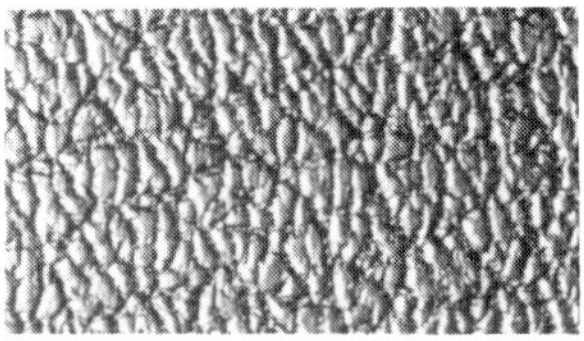

LEATHERETTE

SAMWORTH BOOK BINDING CLOTH TEXTURES

FOUR BINDING CLOTHS ON ONE TITLE
from left to right:
Pebble Grain, Leatherette, Smooth Cloth, Woven Silk

This is but one example of the number of variants found on Small Arms Technical Publishing Co. titles.

Other standard books for riflemen, handgun marksmen, sportsmen and all users of firearms, which have also been gotten out by the publishers of this book, are:—

A RIFLEMAN WENT TO WAR, by **Capt. Herbert W. McBride.** A splendid narrative of the author's personal experiences and observations in the World War. All about sniping and snipers, their equipment and its use, battle firing and the training of the individual rifleshot for war. 400 pages. $3.50

TEXTBOOK OF FIREARMS INVESTIGATION, IDENTIFICATION & EVIDENCE, by **Lt. Col. Julian S. Hatcher.** The most complete and applicable work in any language. Relates to the thorough investigation of crimes committed with firearms, their legal identification and the preparation and presentation of evidence in court trials. For use of police, peace officers, prosecuting attorneys, investigators, members of the legal profession and the judiciary. Some 900 pages with 300 illustrations. $7.50

TEXTBOOK OF PISTOLS AND REVOLVERS, by **Lt. Col. Julian S. Hatcher.** The only *complete* book ever written for handgun users. All about *all* modern pistols and revolvers—their ammunition—their ballistics—their use. 532 pages, 190 illustrations. $4.25

BOOK OF THE SPRINGFIELD, by **Capt. Edward C. Crossman.** A treatise on the exceedingly popular .30 '06 Springfield rifle in all its forms—military—target—sporter. Also its ammunition in target and hunting types and the sights applicable for use with it—both telescopic and metallic. 450 pages and illustrations. $4.00

MILITARY AND SPORTING RIFLE SHOOTING, by **Capt. E. C. Crossman.** Modern and complete in all details. Devoted solely to the shooting of modern highpower rifles—military—target—sporting—of all types and calibers. For either "practical" hunter or user of the most elaborate "Free" rifle used in International competitions. 500 pages. 100 illustrations. $4.50

MODERN GUNSMITHING, by **Clyde Baker.** The most popular book ever written for shooting men and the only one which treats the subject of gunsmithing in a *complete yet practical* manner. For either professional or amateur workman. 525 pages. 200 original illustrations. $4.50

.22 CALIBER RIFLE SHOOTING, by **C. S. Landis.** A book for the user of the increasingly popular .22 rim-fire rifle—either small-bore target shot or outdoorsman. Full of practical hunting dope. 400 well illustrated pages. $3.75

MODERN SHOTGUNS AND LOADS, by **Major Charles Askins.** The most complete book available on the construction and use of modern shotguns and their ammunition. Includes a complete course in field shooting of the shotgun. 436 pages. 100 illustrations. $4.00

WILDERNESS HUNTING AND WILDCRAFT, by **Townsend Whelen.** The only book which really teaches the hunting and shooting of our American big game. A great work by a foremost authority. 340 pages. 75 illustrations. $3.75

For more complete description, write for catalog to:

THOMAS G. SAMWORTH
Small-Arms Technical Publishing Co.
Onslow County, N. C.

Other standard books for riflemen, handgun marksmen, sportsmen and all users of firearms, which have also been gotten out by the publishers of this book, are:—

A RIFLEMAN WENT TO WAR, by **Capt. Herbert W. McBride,** A narrative of the author's experiences and observations in the World War. All about sniping and snipers, their equipment and its use, battle firing and training the individual rifleshot. 400 pages. $3.50

TEXTBOOK OF FIREARMS INVESTIGATION, IDENTIFICATION & EVIDENCE, by **Lt. Col. Julian S. Hatcher.** Relates to the thorough investigation of crimes committed with firearms, their legal identification and the preparation and presentation of evidence in court trials. For use of police, peace officers, prosecuting attorneys, investigators, members of the legal profession and the judiciary. Some 900 pages with 300 illustrations. $7.50

TEXTBOOK OF PISTOLS AND REVOLVERS, by **Lt. Col. Julian S. Hatcher.** The only *complete* book ever written for handgun users. All about *all* modern pistols and revolvers—their ammunition—their ballistics—their use. 532 pages, 190 illustrations. $4.25

BOOK OF THE SPRINGFIELD, by **Capt. Edward C. Crossman.** A treatise on the exceedingly popular .30 '06 Springfield rifle in all its forms—military—target—sporter. Also its ammunition in target and hunting types and the sights applicable for use with it— both telescopic and metallic. 450 pages and illustrations. $4.00

MILITARY AND SPORTING RIFLE SHOOTING, by **Capt. E. C. Crossman.** Modern and complete in all details. Devoted solely to the shooting of modern highpower rifles—military—target—sporting—of all types and calibers. For either "practical" hunter or target shot. 500 pages. 100 illustrations. $4.50

MODERN GUNSMITHING, by **Clyde Baker.** The most popular book ever written for shooting men and the only one which treats the subject of gunsmithing in a *complete yet practical* manner. For professional or amateur. 525 pages. 200 illustrations. $4.50

.22 CALIBER RIFLE SHOOTING, by **C. S. Landis.** For the user of the .22 rim-fire rifle—either small-bore target shot or outdoorsman. Full of hunting dope. 400 well illustrated pages. $3.75

MODERN SHOTGUNS AND LOADS, by **Major Charles Askins.** The most complete book available on the construction and use of modern shotguns and their ammunition. Includes a complete course in field shooting. 436 pages. 100 illustrations. $4.00

WILDERNESS HUNTING AND WILDCRAFT, by **Townsend Whelen.** The only book which really teaches the hunting and shooting of our American big game. 340 pages. 75 illustrations. $3.75

For more complete description, write for catalog to:

THOMAS G. SAMWORTH
Small-Arms Technical Publishing Co.
Onslow County, N. C.

DUST JACKET AD VARIANTS

On the left is an early issue Onslow County dust jacket back panel ad, with a later issue jacket ad shown on the right. Both list the same titles, but the specific ad copy is slightly different, and the "Samworth's Books" cartridge case head appears at the top of the later ad. The jackets are from #40, *Telescopic Rifle Sights*.

to differentiate between it and the predecessor covering. Woven silk is just what comes to mind; the embossing lines have a perpendicular, cross-hatched pattern.

Pebble Grain binding cloth employs a fine weave cloth substrate as a carrier for a heavy coating of dyed filler pigment that is stamp- or roll-embossed with a fine, bumpy, pebble-like finish.

Leatherette is an embossed filled cloth with a "cowhide" or "bison" grain. It was used on only a few of the later issues of the first impressions of the 18mo. titles such as *Telescopic Rifle Sights.*

Fine Cloth refers to a dyed, lightly filled or unfilled fine-weave cloth with small diameter threads woven into a high thread per inch count.

Smooth Cloth is a very smooth, heavily filled vellum finish binding cloth and one must look closely to discern the weave of the cloth through the filler. This is the type of binding cloth found on many of today's hardcover books, and was used on several later impressions of SATPCO titles.

BOOK FORMATS AND SIZES

The book trade uses distinctive terminology with respect to book formats, some of which include: 12mo; Duodecimo, sometimes pronounced "twelvemo", 7 1/2" X 4 1/2"; 8vo, or "Octavo", ranging from "Crown Octavo" or 7 1/2" X 5" to "Royal Octavo" or 9-1/2" X 6 1/2", and Imperial Octavo, 11 1/2" X 8 1/2".

These names arise from the Latin words for numbers, and in the strict sense, refer to the number of times a printed sheet is folded to yield pages. If one takes a large sheet of paper and folds it three times and consecutively numbers the pages without trimming, the result upon unfolding is the page layout for 16 pages, eight-up. This is octavo format.

Presently, these terms are generally accepted by the used and rare book trade to describe the approximate physical dimensions of books regardless of the number of folds in the signatures therein.

The finished book size depends on the size of the sheet of paper the pages are printed upon. A typical standard sheet size is 25" X 38". Fold this sheet four times and the result is a "32" or 32-page signature, 16-up, with a nominal finished size of 5" X 7-1/2". Only nine such sheets of paper would account for the leaves in this book. It follows that it is more economical to print larger signatures because more pages are printed with each pass through the press, requiring less press time.

All Samworth titles can be classified into five groups according to size and binding style. Each classification has an arbitrary "nickname." For example, "Early Sams" are the octavo or simply "8vo" titles published in Marshallton, Onslow County and Plantersville. Case bindings are typically "silk" in the early issues and "woven silk" or "fine cloth" in the later ones. Both carry a decorative bas-relief embossing on the front cover. Examples include *Modern Shotguns and Loads*, *Modern Gunsmithing*, *.22 Caliber Rifle Shooting*, *Textbook of Pistols and Revolvers* and *Military and Sporting Rifle Shooting.*

The next category encompasses the "Small Sams", the inexpensive, thin 12mo books with Onslow County and Plantersville imprints such as *Big Game Rifles and Cartridges*, *Firearm Blueing and Browning*, *Telescopic Rifle Sights* and *Woodchuck Hunter.* First impressions of these had "pebble grain" binding cloth, which was very fragile in the hinge areas and was changed to the more durable "fine cloth" in subsequent issues, a few of which had "leatherette."

At the other end of the scale, "Deluxe Sams" are the tall 8vo titles of Plantersville and Georgetown origin characterized by gilt lettering overprinted on title blocks of a color complementing and contrasting with that of the case binding. *The American Wild Turkey*, *African Rifles and Cartridges*, and *Textbook of Automatic Pistols* are a few that fall under this heading. Deluxe Sams used buckram in general, but first printings of the books published at Plantersville that introduced

"Early Sam" Class Example

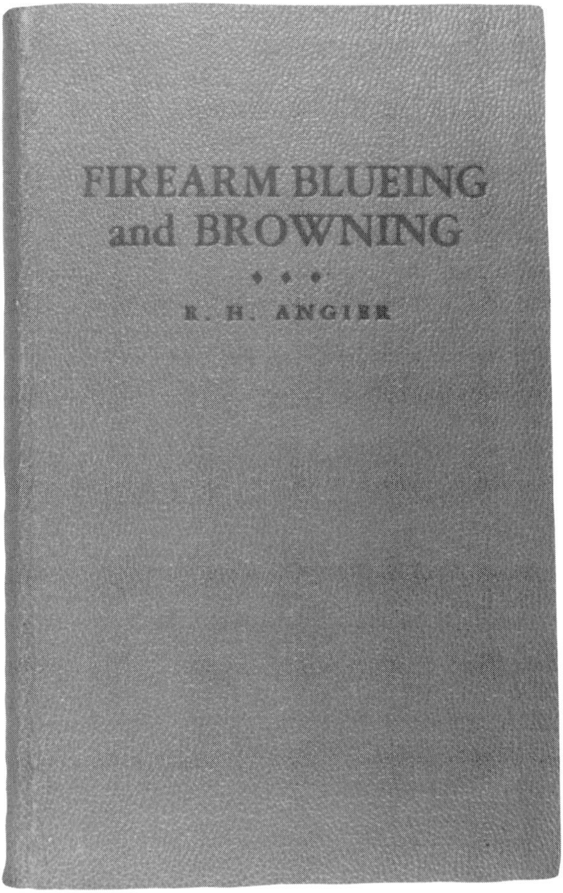

"Small Sam" Class Example

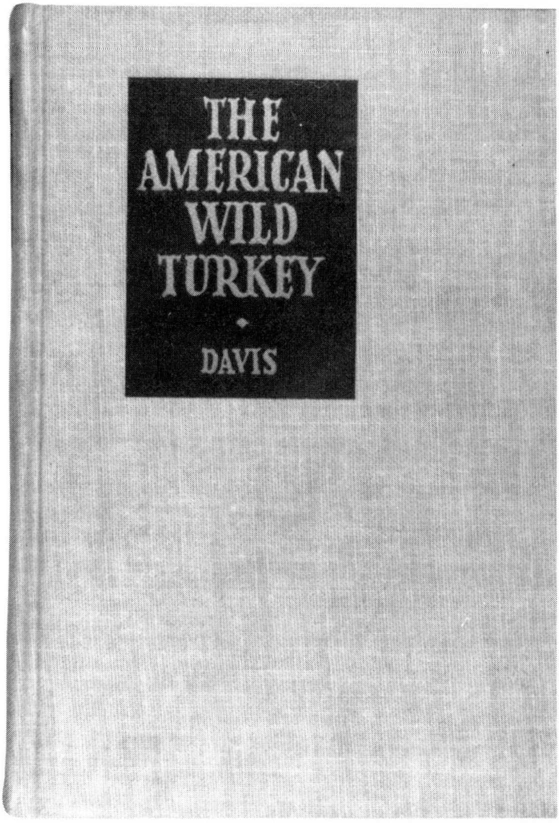

"Deluxe Sam" Class Example

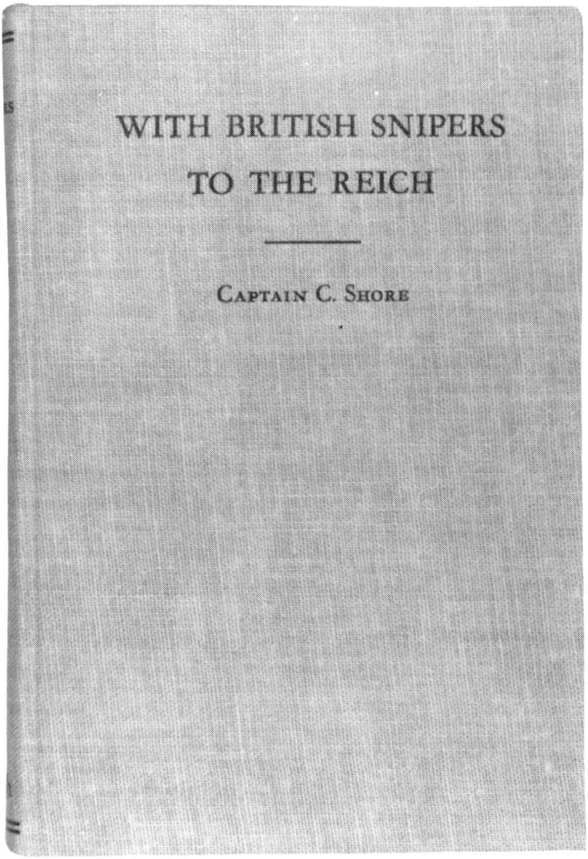

"Standard Sam" Class Example

"F D & A Booklet" Class Example

this format and style are found with woven silk binding cloth covering their boards, *English Pistols and Revolvers* being the best example.

Those books published in Plantersville and Georgetown having coarse weave buckram covers are dubbed "Standard Sams", due to the standard type of binding material that became popular after World War II. The 1951 edition of *Book of the Springfield*, together with *Shots Fired in Anger*, and *Hunting With the Twenty-Two* are examples of this type. The cloth used on this class of Samworth books varies with respect to the degree of pigment filler employed. *Shots Fired in Anger* has a heavily-filled cloth case, whereas the cover on the second edition of *Book of the Springfield* is only slightly filled.

The fifth category contains the "FD&A Booklets" of the *Firearm Design and Assembly* series, each Imperial 8vo in size. The three by Linden have thin wrappers covers, while the Kennedy is bound in stiff boards.

Two titles are unique in format if not binding material; *Handloading Ammunition* was the first Small Arms Technical Publishing Co. book and is a short 8vo volume bound in coarse brown buckram with orange lettering. *Ballistic Records*, the only SATPCO book with Samworth as the author, is almost folio sized in a horizontal format and bound in a tan colored coarse canvas. For classification purposes, these two are included with the "Standard Sams".

IDENTIFICATION OF IMPRESSIONS AND ISSUES

One of the major frustrations in the collecting of Samworth books is the correct identification of a first or later impression of a given title, due to a lack of explicit information. Several clues exist which the informed individual may use to advantage: 1) Copyright date; 2) Location of publication; 3) Title page date, if so marked; 4) Ad page date; 5) Ad page(s) contents; 6) Case binding materials; 7) Dust jacket and its ads, and 8) Independent points and/or provenance.

The copyright date is the primary detail that will assist the collector in the identification of a title. It is used in conjunction with the remaining clues. All Samworths were copyrighted by either Small Arms Technical Publishing Company or Thomas G. Samworth. Earlier titles saw the use of the former in the copyright statement, but as time went on, Samworth copyrighted most of the titles in his name. Because Stackpole Publishing Co. has reprinted several Samworth titles, that name and the Harrisburg, PA address is sometimes found on the title page verso of a Samworth title. Under no circumstances did the words "Stackpole Publishing Co." appear in the copyright statement or title page of any original Samworth-published impression. A Samworth title so marked is always a later Stackpole reprint, done sometime between the late-1950s to the present.

The location of publication is a simple check to determine the impression or issue for several titles. One need only keep in mind that Samworth operated from four separate locations during his publishing career: 1927 to 1932--Marshallton, DE; 1932 to 1940--Onslow County, NC, also referred to as "Marines"; 1941 to early 1943--Plantersville, SC; 1943 to mid-1950s--Georgetown, SC. It should be well noted that he continued to use the Plantersville post office until late 1948, even though he relocated to Dirleton in 1943.

As an example, a title with a copyright date of 1936 such as Reichenbach's *Sixguns and Bullseyes* should be marked as being published in Onslow County if it is a first printing. Plantersville imprints of *Sixguns and Bullseyes* also carry a 1936 copyright, but it is obvious that they must be later issues.

On a few titles, a date appears at the bottom of the title page, presumably indicating the year of issue. As an example, each issue of *Advanced Gunsmithing* has a different title page date; would that there be similar markings on all issues of all titles!

Samworth developed an early habit of advertising other titles on what would normally be the blank rear flyleaves of his books. The practice began with the Marines-Onslow titles, the ad appearing on a single page listing all titles published to date. It should be stressed that the Marshallton books did not have this ad page. Because of the heavy black border around the list of titles, this early type of ad is informally referred to as the "block" ad. The early Marines and Onslow titles, e.g. *The Book of the Springfield*, were published in an style identical to that of the Marshallton books, and carry no ad page date.

The introduction of the "small Sams", the inexpensive $1.50 to $2.00 little manuals that sustained the embryonic publisher through the Depression, saw the ad page date appear with regularity. Almost always found at the lower left corner of the ad page, this date is given as a month and year. Using the *Sixguns and Bullseyes* example above, it was established that a first impression issue must have a 1936 copyright date, be marked Onslow County, and referring to the ad page, have it dated August, 1936.

Every rule has its exceptions, and this axiom pertains also to Samworth books. Some first impressions have a copyright date of one year, and an ad page date of an early month the following year. Therefore, there will not always be agreement between copyright date and ad page date, or in a few cases, ad page date and location.

For example, the compiler has in his collection a first impression of *Ordnance Went Up Front* with the following data: Copyright 1948, Plantersville on the title page verso; March, 1948 ad page date, and the ad page address is Thomas G. Samworth, Georgetown, SC.

Big Game Hunting and Marksmanship—by K. F. Lee	$2.00
Advanced Gunsmithing—by W. F. Vickery	4.00
Elementary Gunsmithing—by Perry D. Frazer	2.00
Handloader's Manual—by Earl Naramore	3.50
Firearm Blueing and Browning—by R. H. Angier	2.50
Sixgun Cartridges and Loads—by Elmer Keith	1.50
Big Game Rifles and Cartridges—by Elmer Keith	1.50
Automatic Pistol Marksmanship—by Wm. Reichenbach	1.50
Sixguns and Bullseyes—by Wm. Reichenbach	1.50
Telescopic Rifle Sights—by Townsend Whelen	1.50
The Woodchuck Hunter—by Paul C. Estey	1.50
English Pistols and Revolvers—by J. N. George	4.00
A Rifleman Went to War—by Capt. Herbert W. McBride	3.50
Textbook of Firearms Identification—by Col. J. S. Hatcher	7.50
Textbook of Pistols and Revolvers—by Col. J. S. Hatcher	4.25
Modern Gunsmithing—by Clyde Baker	4.50
Book of the Springfield—by Capt. E. C. Crossman	4.00
Military & Sporting Rifle Shooting—by Capt. E. C. Crossman	4.50
.22 Caliber Rifle Shooting—by C. S. Landis	3.75
Modern Shotguns and Loads—by Charles Askins	4.00
Wilderness Hunting and Wildcraft—by Townsend Whelen	3.75

In hand, for future publication, are:—

Gunstock Design and Stockmaking—by Alvin Linden
Varmint and Small Game Rifles—by Elmer Keith
Textbook of Automatic Pistols—by Col. R. K. Wilson
English Guns and Rifles—by J. N. George

For catalog, write:—

THOMAS G. SAMWORTH
Plantersville
South Carolina

EARLY PLANTERSVILLE AD PAGE

This ad page dates the book in which is appears by virtue of the fact that the latest book listed is #26, *Big Game Hunting and Marksmanship.* The Wilson book, *Textbook of Automatic Pistols* (43), and George's *English Guns and Rifles* are stated as being in preparation, as are two works by Linden and Keith which were never published.

SAMWORTH BOOKS ON FIREARMS 321

Samworth Books on Firearms

PENNA. RIFLES AND RIFLEMAKERS (In Preparation) — R. M. Smith
HUNTING WITH THE TWENTY-TWO (In Preparation) — C. S. Landis
THE AMERICAN WILD TURKEY — Henry E. Davis
WITH BRITISH SNIPERS TO THE REICH — C. Shore
AFRICAN RIFLES AND CARTRIDGES — John Taylor
ORDNANCE WENT UP FRONT — Roy F. Dunlap
SHOTS FIRED IN ANGER — John B. George
ENGLISH GUNS AND RIFLES — J. Nigel George
ENGLISH PISTOLS AND REVOLVERS — J. Nigel George
TWENTY-TWO CALIBER VARMINT RIFLES — C. S. Landis
PROFESSIONAL GUNSMITHING — Walter J. Howe
MODERN AMERICAN RIFLES, 1890 (Reprint) — A. C. Gould
MODERN AMERICAN PISTOLS, 1892 (Reprint) — A. C. Gould
THE GUNSMITH'S MANUAL, 1882 (Reprint) — Stelle & Harrison
SMALL-ARMS DESIGN AND BALLISTICS, Vol. I — Townsend Whelen
SMALL-ARMS DESIGN AND BALLISTICS, Vol. II — Townsend Whelen
TELESCOPIC RIFLE SIGHTS, (1944) — Townsend Whelen
TEXTBOOK OF AUTOMATIC PISTOLS — R. K. Wilson
A RIFLEMAN WENT TO WAR — H. W. McBride
TEXTBOOK OF FIREARMS IDENTIFICATION — J. S. Hatcher
TEXTBOOK OF PISTOLS AND REVOLVERS — J. S. Hatcher
MODERN GUNSMITHING — Clyde Baker
BOOK OF THE SPRINGFIELD — E. C. Crossman
MILITARY AND SPORTING RIFLE SHOOTING — E. C. Crossman
BIG GAME HUNTING AND MARKSMANSHIP — K. F. Lee
ELEMENTARY GUNSMITHING — Perry D. Fraser
ADVANCED GUNSMITHING — W. F. Vickery
HANDLOADER'S MANUAL — Earl Naramore
FIREARM BLUEING AND BROWNING — R. H. Angier
SIXGUN CARTRIDGES AND LOADS — Elmer Keith
BIG GAME RIFLES AND CARTRIDGES — Elmer Keith
AUTOMATIC PISTOL MARKSMANSHIP — Wm. Reichenbach
SIXGUNS AND BULLSEYES — Wm. Reichenbach
FIREARM DESIGN AND ASSEMBLY — Alvin Linden
  INLETTING THE STOCK BLANK
  SHAPING THE INLETTED BLANK
  FINISHING THE GUNSTOCK

Above are listed the Samworth Books on Firearms now offered for sale or in process of publication by

THOMAS G. SAMWORTH
Small-Arms Technical Publishing Company
Georgetown, South Carolina
U. S. A.

December 1948

GEORGETOWN AD PAGE

This ad page is taken from (7) *The American Wild Turkey*. The ad page date of December, 1948 can be seen in the lower left corner. The Landis *Hunting With the Twenty-Two* (25) and the never-published *Pennsylvania Rifles and Riflemakers* by Smith are listed as being in preparation. Only three *Firearm Design and Assembly* titles are cataloged, Linden having passed away in 1946 before the other three manuscripts in the series could be completed.

Two locations in this one copy might seem suspicious, but serve to identify it as a first impression, as the Georgetown Post Office branch took over servicing Dirleton Plantation late in that year. Again, one should not rely on ad page date alone.

The books advertised on an ad page are another means of identifying a given copy. A true 1936 first impression, first issue of *Sixguns and Bullseyes* would not have post-1936 titles listed on the ad page. As a 1943 impression of that title exists with the ad page so dated, the question of ad page contents becomes moot. There are, however, a few titles with undated ad pages that the enthusiast will be able to correctly identify with a little attention to ad page contents. One example of that is *A Rifleman Went to War*. There are Plantersville impressions that have different case bindings and ad pages listing different titles. The ad page of the latest Plantersville printing gives Lee's *Big Game Hunting and Marksmanship* as the newest book published, making it a 1941 or early 1942 issue. The earlier Plantersville *Rifleman* does not list the Lee book.

Case binding materials were previously described in detail, and can help differentiate a first impression from later ones. This technique is most useful on the "Small Sams" class of books. The first impression of every title in this category used the thin "pebble grain" embossed cloth case. Subsequent printings were usually cased with a thicker, more conventional woven "fine cloth" binding.

The lettering stamps used on the binding can sometimes be used to differentiate between early and late issues. Crossman's *The Book of the Springfield* is represented by examples bearing "STP CO." above the spine tailcap, and later issues having "Samworth"; this latter marking not being introduced until 1938.

The next detail to observe is the dust jacket, if present, but one should be cautioned that the presence of a dust jacket does not necessarily mean that the jacket is a proper one to match the book. Most Marshallton titles and some first impression Onslow books had jackets that were relatively devoid of color; usually one color of ink was used on off-white, beige, or yellow stock. Later Plantersville-imprints had jackets with three or more colors. Crossman's *Military and Sporting Rifle Shooting* and *The Book of the Springfield*, as well as Hatcher's *Textbook of Pistols and Revolvers* are a few examples.

The informed enthusiast can also pay attention to the advertisements on the dust jackets themselves. To illustrate, an Onslow imprinted copy of *Sixguns and Bullseyes* with a 1936 ad page date should not be wrapped in a jacket advertising Lee's *Big Game Hunting and Marksmanship* or Vickery's *Advanced Gunsmithing*, both post-1936 titles.

An unfortunate practice in which many book dealers engage is to switch jackets or sell jackets separately. It is in this manner that books and jackets are improperly wedded. Although the purist may scoff, it could be argued that because Samworth books (and moreso dust jackets) are so scarce relative to other books in general, the presence of a jacket on a book is highly desireable regardless of whether or not the two match. The compiler does not share this opinion.

While some collectors disdain previous owner (p.o.) marks or bookplates, such independent and provenent points like author inscriptions are most useful in determining issue or impression, particularly if the inscriptions are dated.

The collector or dealer will find that time spent studying such details as described above will serve to quickly answer most questions that arise concerning a specimen. The serious collector can thus reasonably ascertain the impression of an item and its suitability for his collection; likewise the dilligent dealer can use the same procedure to correctly differentiate a more valuable Samworth imprint from a later Stackpole reprint.

MISCELLANEOUS DETAILS

Other Samworth "trivia" of which the collector or dealer should be aware include: Only one Samworth title was published with each successive impression noted on the title page verso, and that is *Textbook of Firearms Investigation, Identification, and Evidence* by Hatcher, see #19.

Only three titles had stated second editions: *Modern Gunsmithing* by Baker came in 1928 and 1933 editions so marked, the latter having no change in content with respect to the first.

The second title with two stated editons is Crossman's *The Book of the Springfield*, with editions published in 1932 and 1951. The 1951 edition reprints exactly the Crossman material of the first, with new material by Roy Dunlap at the end of each chapter, serving to update the earlier work.

The third is Whelen's *Telescopic Rifle Sights.* It first appeared in 1936, with a 1944 edition making its debut in that year and advertised as such thereafter. The 1944 book has two chapters added to the complete text found in the original, but only on the title page and table of contents page is attention called to this being a revision or update. Ad pages in other Samworth titles dated 1944 and later sometimes list both editions, with the second edition shown simply as *Telescopic Rifle Sights* (1944); therefore it is not explicitly indicated as a second editon.

Samworth probably did not use the term "second editon" aggressively so that he could make longer use of printing plates and binding embossing stamps. Baker's *Modern Gunsmithing* mentioned above has a "transitional" Onslow impression having a first impression case covering a 1933 edition-marked text. Likewise, a Plantersville printing exists with a slightly darker green binding embossed like the Marshallton and Onslow books. In this manner, Samworth undoubtedly saved some money in the utilization of surplus first impression cases whenever possible. This would be in keeping with his reportedly frugal character.

Close examination of several copies of a given title will show differences in size and thickness dimensions. Using *Modern Gunsmithing* as an example, the thin, earlier impressions have olive or dark green bindings in contrast to later kelly-green cased Plantersville printings, which are almost one and one-half times as thick. The disparity in thickness of the two impressions is not due to different editorial content; it is a result of using a higher caliper or heavier basis weight text stock as mentioned above.

Samworth utilized at least three printing houses during the years of operation. George F. Lasher Co. of Philadelphia printed *Handloading Ammunition,* with the Marshallton and Onslow titles printed by Telegraph Press of Harrisburg, PA.

Plantersville titles saw the use of Telegraph Press and Kingsport Press, Kingsport, TN as printers. The Georgetown books are also a mix of Kingsport and Telegraph manufacture.

Samworth books were illustrated by a number of artists, and some authors provided drawings and sketches for their books. The list of author-illustrators includes Townsend Whelen, Julian Hatcher, J.N. George and J. Randall Mattern. Other titles had illustrations provided by E. Stanley (Ned) Smith, Ray Snow, Philip Plaistridge, Oliver B. Hamilton, John B. Moll, Richard Kroth, Edwin Bender and Gayle Hoskins.

Several dust jackets were adorned by sketches or paintings made by Ray Snow, Alden Turner, Ned Smith and Gayle Hoskins. Smith's colorful jacket illustration on *African Rifles and Cartridges* is simply beautiful, as is Hoskins' jacket artwork on *Hunting With the Twenty-Two*

Early Small Arms Technical Publishing Co. books had a "trademark" of sorts on the title page in the form of a shield. This trademark, created by Townsend Whelen, shows a bear and a rifle separated by the diagonal acronym "S.A.T.P.CO."

Miscellaneous Details

Samworth books published later were often decorated with a simple stylized cartridge case head with the slogans "A Samworth Book" or "Book by Samworth" surrounding the indented cartridge primer.

1) Firearm Blueing and Browning
James A. Rogers Library collection

1) ANGIER, R. H.

FIREARM BLUEING AND BROWNING

c1936 SATPCO Onslow County, N.C. 1936 ad page date. Tall 12mo.

CASE: Orange pebble grain embossed thin cloth. Blue lettering on front and spine. All edges orange ink-washed. Two-color blue and brown dust jacket illustrated with a gunsmith blueing shotgun barrels.

CONTENTS: 151 pp. Seven chapters incl: Introduction; Browning Process; Preparation of Blueing and Browning Solutions; Brownes (Blues) and working instructions; Pickling, etching, matting. Finishing twist barrels; Derusting and Denickling; Phosphatising (rustproofing). Bibliography and Appedix follow.

Frontis. of technician blueing a rifle barrel. Contains dozens of formulae for the finishing of firearms, with detailed instructions for the correct preparation and application steps required.

SAMWORTH PROMOTIONAL: "Here is a qualified, practical and complete treatise covering the art--some 300 formulae for various steels and processes, together with complete working instructions which clearly explain how a gun barrel or other parts may be oxidized with solutions compounded at home. These solutions range from one made from common table salt, on up to the most involved and high sounding mixture of eight or ten chemicals--and some of the more simple ones give results equal to those obtained with the most complicated solution in the book.

"There are given operating directions and formulae for all the various processes--hot blueing--20 minute solution--temper blueing--slow rusting--one application--commercial methods for small parts--chemical removal of rust, old blueing, nickle plating--

chemical matting and etching, pickling--etc. All steels and gunmetals, from early soft carbon to modern 'rustless' are treated. Applicable by either amateur or professional gunsmiths and is a manual of considerable practical value to workers in commercial metals, welders, and repairmen of automotive and other steel equipment. 155 pages."

SUBSEQUENT PRINTINGS AND ISSUES:

c1936 SATPCO Plantersville, SC. 1941 ad page date. 12mo. Case is similar to that of the first above, but with thicker boards covered with a fine weave cloth, either orange or green in color. Dust jacket is closely similar to that of the first impression.

c1936 SATPCO Plantersville, SC. 1945 ad page date. 12mo. Navy and dark brown fine weave cloth cases have been seen, suggesting two or more issues of the 1945 impression. Text paper stock is bluish-gray and not as smooth as that in earlier issues. Dust jacket is closely similar to that of the first impression.

c1936 SATPCO Georgetown, SC. 1950 ad page date. 12mo. Dark brown heavily filled smooth cloth case binding. Dust jacket is illustrated similarly to that of the first with a sketch of a gunsmith blueing a barrel, but a close comparison shows slightly different artwork suggesting a new illustration. Additionally, it is printed in a single color of dark blue.

SUBSEQUENT EDITIONS:

Various Stackpole facsimile reprint issues exist, so marked, with dark brown bindings and single-color dust jackets illustrated like that of the 1950 Samworth issue.

NOTES: A title that has stood the test of time; it is still in print by Stackpole. The dozens of formulae compiled by author Angier represent a "cookbook" filled with data that are likely not available in any other source today. Those wishing to make use of the various solutions may find it difficult to identify some of the ingredients; in some formulae they are given by their popular and colloquial names.

The compiler has been advised that copies exist with green cases, but he has not personally examined any. First impression Samworth specimens are somewhat hard to find, especially with dust jacket. The compiler attributes this to the book being a "working" title; one that spends most of its time open on the bench, being marked by yet more stains and layers of oil and dirt.

The thin pebble grain embossed boards are very fragile; with some working, the embossing readily cracks and flakes off the hinge areas, exposing the cloth underneath. A later issue is recommended as a "reading" copy to preserve the condition of a first impression in one's collection.

2) Modern Shotguns and Loads
M. L. Biscotti collection

2) ASKINS, Capt. Charles (Sr.)

MODERN SHOTGUNS AND LOADS together with a treatise on the art of wing shooting

c1927 SATPCO Marshallton, DE. No ad page. Tall 8vo.

CASE: Light putty or ivory colored fine-line embossed cloth. Gilt lettering on front and spine, decorative gilt head- and tailstriping. "STP CO" in gilt above tailcap. Front cover has decorative octagonal-framed embossing of two smoking shotgun hulls in grass. Dust jacket illustration showing a hunter aiming a shotgun at birds in flight; one is falling amid loose feathers.

CONTENTS: 416 pp. Eighteen chapters (282 pp.) in Part I incl: A Retrospect; Double Mechanisms; Special Mechanisms; Double Shotgun Construction and Finish; Engraving; Stocks and Stocking; Utilities, Fads, and Fancies; Choke Boring; (Continued); Gages and Their Use; Boring and Barrel Length; Patterns; Pellets per Load; Popular Shot Sizes; Factory Shot Shell Loading; Factory Ballistics; Testing Factory Ammunition; Ammunition Performance at Long Range; Velocity, Energy, Pattern and Range; Range of Small Bores and Small Shot; Range Based on Patterns; Pattern Tests on Game; Fitting the Cartridge to the Gun; Smokeless Shotgun Powders; Shot Stringing; Handloading Shotgun Shells; Special Handloading Problems; Care and Cleaning of Shotguns; Glossary; Facts and Figures.

Fifteen chapters (121 pp.) in Part II incl: An Essay on Wing Shooting; Aiming; Lead; Swinging and Snapping; Correct Positions; Mechanical Execution; Class, Form, Style and Time; Psychology of Wing Shooting; Flinching; Hitting vs. Missing; Using the Second Barrel; What Constitutes Good Shooting?; The Boy and His Gun; The Woman Afield; Free Shooting; Index.

Frontis. of four views of an engraved Parker Bros. double, text illustrated with photos of contemporary double, pump and autoloading shotguns, wingshooting form and positions, plus drawings of shot patterns, charts of loads, ranges, etc.

SAMWORTH PROMOTIONAL: "Captain Askins has here written the latest and most comprehensive work on the construction and possibilities of American Shotguns; their ammunition and their use in the hunting field. Actually, he has given you two books combined in one as the second part of the most practical work covers "The Art of Wingshooting." Part I covers all the technical and ballistic properties of modern shotguns, boring, fitting, pattern requirements and analysis and ammunition performances. Outstanding chapters are those showing killing energy of all the popular loads at different ranges and on all kinds of feathered game. In the wingshooting part Askins gives his experience gained in fifty years intensive hunting with the shotgun. This book will teach you how to become a good wingshot on all kinds of game; how to aim, lead, swing or snap and how to use the second barrel; all taught by a most accurate and practical shooter. Bound in our standard size and style of silk cloth, 436 pages of reading matter with 100 illustrations."

SUBSEQUENT PRINTINGS AND ISSUES: None.

SUBSEQUENT EDITIONS: None.

NOTES: Due to the light color of the case, this title is rarely seen in Near Fine or better condition. Most copies are extremely soiled. As with all decoratively-embossed early Samworths, the cover embossing generally "welts" the endpapers, flyleaves, frontis, and the first few pages.

This exhaustive work eventually came to be considered a definitive text on shotguns for decades, serving as the basis for later books by others. Ironically, it was not a big seller for Samworth when introduced, as he related in a 1931 correspondence:

"(The Askins book is) a complete flop and right now I am about $1800 in the hole on account of it. Despite this I have paid Cap $350 and will send him more until I pay him $1000 as I guaranteed him this amount. I will just about come out on this book in eight more years."

Nonetheless, the thorough, well written text explains covered topics in great detail; excellent information is presented on wingshooting techniques. This is a reference of extreme interest to every present-day shotgun enthusiast, as well as fans of the elder Charles Askins. Samworth's comments shed some light on the reason for this title's relative obscurity; the single print run consisted of approximately 2600 copies, making this title one of the rarest Samworth books. This fact is not reflected in the pricing levels seen by the compiler. Advertised copies seem to be priced with respect to overall intrinsic content.

3) Modern Gunsmithing
1928 Edition
B. R. Smith collection

3) BAKER, Clyde

MODERN GUNSMITHING A Manual of Firearms Design, Construction and Remodeling, for Amateurs and Professionals

In two stated editions: c1928, Marshallton, DE; c1933 Onslow County, NC.

FIRST EDITION:

c1928 SATPCO Marshallton, DE. Undated ad page. Short crown 8vo.

CASE: Dark olive green silkweave embossed cloth. Gilt lettering on front and spine. Decorative gilt head- and tailstriping. "STP Co." above tailcap. Decorative embossing of a pair of hands checkering the grip of a rifle stock. Dust jacket illustrated with a sketch in single color of a gunsmith apparently draw-filing a rifle barrel in a vise.

CONTENTS: 530 pp. Introduction and thirty-four chapters incl: Home Gunsmithing; The Gun-crank's Workshop; Tooling Up; Special and Home Made Tools and Equipment; Materials and Metals Necessary; First Steps for Beginners; General Shop Practice and Use of Tools; Wood for Gunstocks; Gunstock Design; Stockmaking: Laying Out and Inletting; Stockmaking: Shaping and Fitting; Checking and Carving; Finishing and Polishing Stocks; Repairing and Remodeling Stocks; Rifle Barrel Design and Fitting (by Townsend Whelen); Chambering and Barrel Work (by Townsend Whelen); Cartridge Design and Manufacture (by Townsend Whelen); Striking and Polishing Barrels; Engraving and Polishing of Metal; Blueing and Browning; Annealing, Hardening and Tempering; Case Hardening; Soldering, Brazing and Welding; Manufacture and Substitution of Small Parts; Hard Fitting of Small Parts; Cleaning Bores and Removing

Obstructions; Lapping Barrels and Bore Polishing; Adjusting Trigger Pulls and Actions; Fitting Sights and Telescopes; Remodeling Military and Obsolete Rifles; Shotgun Repairs and Alterations; Pistols and Revolvers; Restoration and Repair of Old Firearms; Emergency and Field Repairs. Appendices follow.

Frontis. of an equipped workbench; illustrated with 200 sketches and halftones.

SAMWORTH PROMOTIONAL: "This standard work contains thirty-four chapters, occupying 525 pages of closely set type, and has more than 200 original illustrations from actual photographs and line drawings. These chapters include discussions of tool equipment, various woods suited to gunstocks, the design and construction of modern gunstocks, how to fit the individual shooter, polishing metal parts, blueing, browning and coloring of metals (36 proven formulas), welding, soldering and brazing, annealing, hardening, tempering and case-hardening, sights and telescopes and methods of fitting, manufacture and substitution of small parts, selection of tools or the making of special tools as needed, and many other subjects never before covered in print. There are special and extensive chapters on remodeling military and obsolete arms, on shotgun work, revolver and pistol work, reducing trigger pulls, regulating and speeding up actions, etc.

Baker has treated his subject with special reference to the shooter 'back of beyond'--the man in the woods or mountains, or by the boy on the farm or ranch--located far from supply shops, stores or gunsmiths. But the latter have also been constantly kept in mind, and 'Modern Gunsmithing' is all its name implies: a practical, thorough, and up-to-date manual and textbook for either the professional gunsmith or the dealer in firearms who is required to give 'service' with the weapon he sells. The individual shooter who has neither time nor inclination for working over his firearms will find it to be an interesting and valuable treatise on modern firearms and a useful addition to his library; a book full of practical information and data on shooting."

SUBSEQUENT PRINTINGS AND ISSUES,
1ST EDITION: None known.

SECOND EDITION:

c1933 Onslow County, NC. 1933 title page date. No ad page date. Short crown 8vo. Case and dust jacket identical to that of the first edition.

SUBSEQUENT PRINTINGS AND ISSUES,
2ND EDITION:

c1933 SATPCO Plantersville, SC. No ad page date. Short crown 8vo. Estimated to have been printed about 1941, ref. other titles in ad. Case is embossed similarly to the first and early second editions, but is a dark forest green. Dust jacket is virtually identical to that of the first edition.

c1933 SATPCO Plantersville, SC. No ad page date. Short crown 8vo. Believed to have been printed during WWII, judging by the rougher, uncalendered paper stock used for the text. Case is embossed similarly to the previous works, but is a lighter medium green in color. This impression is nearly one-third thicker than previous copies, due to heavier basis weight text stock.

c1933 SATPCO Plantersville, SC. 1950 ad page date. Georgetown ad page address. Crown 8vo. Case is unembossed dark mottled red coarse buckram. Dust jacket illustration is similar to those of the first and early second editions, but a side by side comparison shows they differ in detail and are in fact two unique illustrations.

3) Modern Gunsmithing
1933 Edition
B. R. Smith collection

c1933 Plantersville, SC. 1950 ad page date. Georgetown ad page address. Crown 8vo. Later issue. Case is unembossed oyster white coarse buckram. Dust jacket illustrated as that immediately above, seen in both red and green with black on oyster white stock.

SUBSEQUENT EDITIONS: Various Stackpole facsimile reprints exist, so marked.

NOTES: This was a standard seller for Samworth over the years, and for no small reason. Baker was inspired to write the book by his friend Townsend Whelen, who contributed three chapters. Whelen was dissatisfied with his own 1925 *Amateur Gunsmithing*, a small collection of methods published by the NRA as its first hardcover venture, under the direction of the then-editor of the American Rifleman, Thomas. G. Samworth.

Regarding content, there is no difference between the first and second editions except for minor changes to the material in the appendix charts. For all intents and purposes, the so-called "second editions" should be considered later impressions. Samworth was not liberal with bringing out stated later editions of his books. This is one of only three titles that ever bore the words "second edition."

Samworth had this to say regarding *Modern Gunsmithing* in an early correspondence:

"The Baker book is our best seller in spite of the fact that only Jesus had ever heard of the author before it came out. Almost sold out and Baker is now SUPPOSED to be writing two more books to take its place. To date I have paid him just $2100 in cash as his share and he has sold about $800 worth of my books with no returns; still, genius always was peculiar."

The promised "two more books to take its place" never materialized, and instead of merely issuing a second impression, Samworth chose to market the second "edition" as noted above. Samworth wrote of the second edition in a cover letter introducing a 1933 SATPCO brochure:

"We wish to advise you and your friends who shoot of the publication of a second edition of the most popular book ever written for the shooting fraternity: MODERN GUNSMITHING by Clyde Baker. The first edition of this excellent textbook was sold out some two years ago and since then we have been endeavoring to have written an entirely new volume on the subject. So many delays have occurred in this attempt and there is such a demand for the old work that we have republished the original book. This second edition is an exact reprint and has not been changed in any manner."

The 1933 second edition is represented by Onslow County and Plantersville imprints, with some Plantersville issues having a Georgetown address on the ad page. Early copies of the second edition are so stated on the title page. This is not the case with the later Plantersville and Plantersville/Georgetown issues. The second edition was reprinted by Stackpole Publishing after Samworth ended his publishing career in the mid-1950's, copies of which are so marked, but the dust jackets often say "A Samworth Book On Firearms."

Although written with a somewhat novice or elementary approach, Modern Gunsmithing is extremely detailed and thorough, and best suited to the amateur and intermediate gunsmith. Advanced gunsmiths will find much useful information here, but other works may be more to their interest. It was reprinted so many times by Samworth alone that several variations of binding and dust jacket exist, and the law of averages predicts there are more than those described above.

The author for years operated Baker Arms and Tool Works on Sixth Avenue in Kansas City, MO, contrary to Samworth's perception of Baker's obscurity.

Publishers of
Books of Interest to Riflemen
Revolver and Pistol Shots
Handloaders of Ammunition
Amateur Gunsmiths
Big Game Hunters and
Sportsmen and Marksmen in
General

THE SMALL-ARMS TECHNICAL PUBLISHING COMPANY

MARINES, ONSLOW COUNTY
NORTH CAROLINA

Dealers in
Books of Any Description
Relating to These
Classes of Readers
[Express and Freight Address
Jacksonville, North Carolina]

THOMAS G. SAMWORTH
Manager

April 1, 1933.

Dear Sir:—

We wish to advise you and your friends who shoot of the publication of a second edition of the most popular book ever written for the shooting fraternity:—MODERN GUNSMITHING, by Clyde Baker. The first edition of this excellent text-book was sold out some two years ago and since then we have been endeavoring to have written an entirely new volume on the subject. So many delays have occurred in this attempt and there is such a demand for the old work that we have republished the original book. This second edition is an exact reprint and has not been changed in any manner.

For the man who can really shoot this is the age of made-to-order, custom built weapons; made by expert craftsmen—modern gunsmiths—made with extreme accuracy, of modern materials, dimensioned to fit their owner, adjusted and regulated with the greatest nicety. The making or assembling of such weapons is a science, and this science of modern firearms construction and designing is for the first time given to the shooting world by the one man who knows it thoroughly from the standpoint of the skilled craftsman, the expert marksman and the lifelong practical shooter.

MODERN GUNSMITHING, by Clyde Baker, is a book of 525 pages and 200 illustrations; dealing with the designing, making, remodeling and fitting of ultra-modern rifles, shotguns and pistols. It is a volume for either amateur or professional gunsmiths, for gun-cranks, expert marksmen and sportsmen. It tells how modern weapons should be designed and built in every detail. It goes into each subject so thoroughly that even a novice can, by its use, develop the skill necessary to assemble modern weapons for himself and friends—weapons whose finished value would run into the hundreds of dollars—for the price only of good raw material, a few tools, a little labor and a deal of patience, care and practice.

Though written with special thought to the needs of the gun crank who likes to "tinker" on his guns in a home workshop, MODERN GUNSMITHING will prove no less valuable to the experienced gunsmith, in giving him detailed instructions or formulas for numerous jobs with which he may be thoroughly familiar, particularly in the remodeling of military arms to conform to present day shooter's ideas, for hunting and target work.

Baker's book will open up an entirely new world for the vast army of gun users. It will tell you how you can have built, or how you can construct for yourself a rifle or shotgun dimensioned to your individual build and characteristics; fitted with any combination of sights which YOU want and of a calibre and barrel length suited to yourself. No matter if you never intend to build or remodel firearms yourself, you need this book. It tells you what a modern, efficient and safe weapon should be, it tells you how your gunsmith should build it, it tells you of the many things to beware of in having your weapons built, remodeled or repaired by some of the many locksmiths or garage mechanics who masquerade as gunsmiths. It is a book full of shooting dope, and it tells everything in an interesting and practical way, being full of anecdotes, experiences and humor.

4) Small Bore Rifle Shooting
B. R. Smith collection

4) CROSSMAN, Captain Edward C.

SMALL BORE RIFLE SHOOTING A practical handbook covering in complete detail the new sport of shooting with .22 caliber rifles; equipment, ranges, marksmanship, competitions and clubs.

c1927 SATPCO Marshallton, DE. No ad page. Tall 8vo.

CASE: Dark forest green silk embossed filled cloth. Gilt lettering on front and spine. Decorative gilt head- and tailstriping. "STP CO." in gilt above tail-cap. Front has decorative embossing of a target bulls-eye punctured by bullet holes. Dust jacket unknown.

CONTENTS: 359 pp. Foreword plus seventeen chapters incl: The Game of the Small Bore; the Small Bore Rifle; Ammunition; Small Bore Ballistics; Targets and Ranges; Sights and Their Adjustment; The Telescope Sight; The Rifleman's Equipment; At the Firing Point; Making High Scores; The Care of the Rifle; The Rifle Team; The Rifle Club; The Riflewoman; Some Hints for the Beginner; General Small-Bore Rules and Conditions; Small-Bore History, after which follows "The Small-Bore Rifleman's Directory", a two-page appendix of shooting equipment suppliers.

Frontis. of Sea Girt, NJ, location of shooting matches during the early 20th century. Illustrated with dozens of photos and graphs.

SAMWORTH PROMOTIONAL: "Modern, high-power rifle shooting is expensive, entailing costly ammunition, long trips to special ranges or localities and often heavy range fees. Few can afford enough of it to become really good shots. But small bore rifle shooting--that is, with man-sized, fully-equipped rifles using the .22 long rifle cartridge at ranges of 25 to 200 yards--is within the purse and convenience of anyone. Besides being a splendid sport of itself, small-bore shooting gives that basic training which enables anyone to shoot any rifle expertly. The sport of small-bore shooting is rapidly sweeping this country as it has England. Thousands who can afford practice in no other way are participating. Local, national, and international matches are held annually. In settled parts of America it is the one and only way for the average citizen to become a nail-driving rifle shot.

"Captain Crossman is one of the world's finest riflemen, and was one of the pioneers in small-bore shooting in America. In his book he tells all about it. The book is a practical manual for the small-bore shooter, and the text will also be found to be a liberal education to anyone who uses any rifle for any purpose. There are many principles, data, ballistics and plain truths set forth which will not be found anywhere else and without a knowledge of which any rifleman is handicapped. The book is full of personal anecdotes, experiences and principles, all set forth with a dash of humor. There is not a dry page in it. The authoritative and only work on this subject. 350 pages, 100 illustrations. Handsomely bound in cloth."

SUBSEQUENT PRINTINGS AND ISSUES: Total print run for this title was 2900-odd copies, of which 2000 were bound at first. The remaining copies represent a second issue, but they are indistinguishable from the first issue described above. This text rapidly sold out and was replaced in the SATPCO line by Landis' *.22 Caliber Rifle Shooting.*

The ad pages in some Plantersville titles mention a later printing being on press. Therefore, a second impression may exist, but is unconfirmed at present.

SUBSEQUENT EDITIONS: None.

NOTES: First major work by a contemporary expert, this book details for the reader all points of target shooting with the .22. Information is given on the rifles, e.g. M1922 Springfield, M52 Winchester, Savage 1919, Stevens 414 Single-shot, and B.S.A. Martini Match rifle; target ranges are described as are types and makes of ammunition, telescopic sights, and auxiliary equipment. Shooting positions, equipment, and techniques are thoroughly presented. This book covers exclusively target shooting with the .22; no material on hunting is included.

Samworth was pleased with the outcome of this work: "(Small Bore Rifle Shooting is a) big success, almost sold out (as of 1931) and it is a money maker due mostly to the pronounced wisdom of the writer in grabbing the $250 cash offer instead of any share proposition, Gawd bless every bone in his head."

Only one printing has been confirmed to date, on the basis of correspondence between Samworth and the printer. It was eventually eclipsed by the introduction of #23, *.22 Caliber Rifle Shooting* by Judge Charles S. Landis in 1932. Ad pages in a few Plantersville titles mention *Small Bore Rifle Shooting* as "on the press" in 1944, but this is unsubstantiated at present. If this later impression does exist, the case would almost certainly be woven silk, fine cloth, or smooth cloth, and it would lack the decorative embossing of the target found on the Marshallton book.

The compiler has never seen a dust jacket for this title, despite extensive research among collectors, libraries and dealers to obtain one for photographing.

Based on the sales of this book, Crossman was commissioned to write a corresponding text on centerfire rifle shooting, which evolved into *The Book of the Springfield* and *Military and Sporting Rifle Shooting*, due to the volume of material. See #5 and #6.

5) The Book of the Springfield
1932 Edition
M. L. Biscotti collection

5) CROSSMAN, Capt. Edward C.

THE BOOK OF THE SPRINGFIELD A Textbook Covering all the Various Military and Sporting Rifles Chambered for the Caliber .30 Model 1906 Cartridge; Their Metallic and Telescopic Sights and the Ammunition suited to them.

In two editions: c1932 Onslow County, NC; c1951 Georgetown, SC

FIRST EDITION:

c1932 SATPCO Marines, Onslow County, NC. No ad page date. Tall 8vo.

CASE: Red silk embossed filled cloth. Gilt lettering on front and spine. Decorative gilt head- and tail-striping. "STP CO" in gilt above tailcap. Front has decorative embossing of soldier firing a M1903 Springfield. Dust jacket illustrated with two soldiers firing M1903 rifles at targets on a range.

CONTENTS: 451 pp. Foreword and seventeen chapters incl: The Military Rifle; The Sporting Springfield; The Commercial Sporters; Match and Free Rifles; Military Rifle Sights; Match Rifle Sights; Sights for Hunting; The Target Telescope Sight; The Telescope Sight for Hunting; Telescopic Sights for War; Accuracy and Safety Adjustments; Ammunition Components; Military Rifle Ammunition; Match Ammunition; Sporting Rifle Ammunition; Corrosion and Rifle Cleaning; Ballistics and the Law.

Frontis. of the author working in his laboratory. Illustrated with 100 photographs and charts.

SAMWORTH PROMOTIONAL, FIRST EDITION: "A textbook devoted to all the details and uses of the most popular rifle and cartridge the world has ever known, the 30 Model 1906 Springfield. Contains full data about the many different military, commercial and private makes of rifles for the cartridge together with the very latest information relative to their sights both metallic and telescopic. Both rifles and ammunition are considered from the military, target and sporting standpoint. The data on cartridges is exceptionally complete and is written from military, match and sporting ammunition standpoints. Not only are all the various combinations loaded into .30'06 ammunition analyzed but special illustrations have been made showing sectioned bullets with full details of construction. This data is invaluable to the big game hunter and scientific rifleman. Any user of .30'06 ammunition will find this volume of greatest value. It is also essential to all students of ballistics. 450 pages of the 'dope' bound in our standard silk cloth. Many illustrations."

SUBSEQUENT PRINTINGS AND ISSUES:

c1932 SATPCO Marines, Onslow County, NC. Red-orange woven silk embossed filled cloth with the decorative bas-relief embossing of the first printing. Gilt lettering on front and spine, with "Samworth" in gilt lettering above tailcap. No ad page date. Tall 8vo. Dust jacket illustration is the same as that of the first printing, but three ink colors are used on white paper stock instead of black ink on yellow stock. The decorative front cover embossing of the soldier firing the rifle is not as distinct on this impression as it is on the first, due to the rougher "crosshatched" woven texture of the cover cloth used on this later printing. The texts of both impressions are identical in all respects. The compiler believes this later impression to be a late Onslow or early Plantersville printing due to the binding material and the presence of the gilt "Samworth" at the bottom of the spine. Samworth did not start using his name on the spine until 1938.

SECOND EDITION:

c1951 Thomas G. Samworth Georgetown, SC. 1951 ad page date. Crown 8vo.

CASE: Pale yellow lightly-filled coarse buckram. Brown lettering on front and spine. Decorative brown ink head- and tailstriping. "SAMWORTH" across bottom of spine. Lacks decorative embossing of the first edition. Dust jacket illustration of two Army riflemen firing 1903 Springfield rifles is nearly identical to that of the 1932 edition, differing only in small details.

CONTENTS: 567 pp. Contains a facsimile reproduction of the first edition text in its entirety, with 160 pp. of additional material by gunsmith Roy Dunlap as addenda to each of the original chapters, updating the information in each to 1952. Printed on coated stock.

Illustrated with the B&W halftones and charts of the first edition, plus additional photographs to support the Dunlap material.

SAMWORTH PROMOTIONAL, SECOND EDITION:
"If you are a bolt-action rifleman--and most of us are--you will need this book in your library. Although written around the 1903 Springfield rifle and its 1906 cartridge, this is the most complete and attractively written work on modern rifles and their ballistics that is available today. Its text and teachings will stand for another generation or so. *Book of the Springfield* was originally written by the late Captain Edward C. Crossman, the foremost writer-rifleman of his day and an author whose style of writing has never been surpassed. The first edition was published in 1931 (sic) and has been out of print for several years. Continued and insistent demand has necessitated reprinting in toto the original material compiled by Ned Crossman plus some 160 additional

5) The Book of the Springfield
1951 Edition
M. L. Biscotti collection

pages of text written by Roy F. Dunlap, a firearms authority and writer of ability equal to that of the original author. His new material brings all subject matter up to the year 1952. This 1952 edition is a volume of 567 pages, well illustrated throughout. It is the most exhaustive and thorough work in existence on modern bolt-action rifles--military, target and sporting models. The 16 lengthy chapters differentiate at length between the rifle--the cartridge--sights--and on technical tolerances, adjustments, and care. Its rifle text consists of some 153 pages covering the various military, target and sporting models. The chapters on metallic sights comprise 65 pages--treating fully all types of military, target and hunting sights.

"When it comes to the subject of the modern telescopic rifle sight there is actually a book in itself in the 152 pages in three chapters of up-to-the-minute dope on military, target, hunting and varmint scopes--with special emphasis placed upon these last two types. Included also are comprehensive instructions for the proper fitting and adjustment of the various makes and models. The practicing gunsmith will find this material of particular value in his daily work. To the custom gunmaker it will prove a MUST work of reference. The ammunition section is particularly complete; military, target and hunting ammunition is fully analyzed and discussed, with many illustrations of sectioned bullets included for classification and study.

"Roy Dunlap took up where Crossman left off and their combined effort is a work no rifleman or ballistic student can afford to pass up. The hundreds of thousands of owners and users of Springfield rifles will find in this book everything that is today known in practical experience and ballistic science relative to .30'06 rifle and cartridge matters. *The Book of the Springfield* has long been one of the classics in firearms literature and this revised and greatly enlarged edition will assure its top position for years to come."

SUBSEQUENT PRINTINGS AND ISSUES:

c1951 Thomas G. Samworth Georgetown, SC. 1955 ad page date. Crown 8vo. Later issue. Case and dust jacket identical to that of the 1951 printing.

SUBSEQUENT EDITIONS: Reprinted by Stackpole in the 1960s.

Advertised as being available in 1990, facsimile reprinted in deluxe quarter-leather-binding by Wolfe Publishing Co. in the Wolfe Library Classics series.

NOTES: Long the standard reference for shooters of the Springfield family, this book got its start as a request by Thomas G. Samworth for Crossman to write a text on centerfire rifle shooting. Military, hunting, and target aspects had to be handled, with regard to the many popular cartridges available. The amount of material grew so large that two books were needed to cover it all. *The Book of the Springfield* is concerned with the M1903 rifle and its cousins, the arsenal sporting and target variants, and the commercial sporters. Other .30'06 rifles are mentioned, such as the M1917 Enfield.

The second book, *Military and Sporting Rifle Shooting*, parallels the Springfield text, treating the use of the rifle in a general sense with respect to target, military and hunting uses. These texts have been considered to be companion volumes since their publication. Refer to #6.

The 1951 second edition of *The Book of the Springfield* is a complete reprint of the first edition. Dunlap's addenda are concerned mainly with the subsequent Springfield variations e.g. M1903A3, remodeling Springfields, developments in ammunition and accessories, and information regarding the so-called "low serial number--high serial number" receiver hardness controversy.

It is not particularly scarce, since about 10,000 copies were printed, but demand is always steady.

The present-day Springfield collector will find variants and related equipment are better served by Campbell's *The 1903 Springfields* and Brophy's *The Springfield 1903 Rifles.*

The first edition is somewhat difficult to find with the pale yellow dust jacket in good condition. Also, as with all the early Samworths having embossed covers, the front end papers, flyleaves and frontispiece are often "welted" from the cover embossing transferring through. Later impressions of the first edition have multicolor dust jackets with the same illustration as that of the first.

This title is one of the three Samworths to have a stated second edition, Baker's *Modern Gunsmithing* (#3) and Whelen's *Telescopic Rifle Sights* (#40) being the others.

6) Military and Sporting Rifle Shooting
B. R. Smith collection

6) CROSSMAN, Capt. Edward C.

MILITARY AND SPORTING RIFLE SHOOTING
A complete and practical treatise covering the use of modern military, target and sporting rifles.

c1932 SATPCO Marines, Onslow County, NC. No ad page date. Tall 8vo.

CASE: Dark kelly-green silk embossed filled cloth. Gilt lettering on front and spine. Decorative gilt head- and tailstriping. "STP CO." in gilt above tailcap. Front has decorative embossing of a soldier taking aim with a M1903 rifle. Dust jacket illustration in brown ink on white stock of cowboy taking aim from the sitting position with a rifle while his horse stands nearby.

CONTENTS: 499 pp. Foreword, Introduction and twenty-two chapters incl: A Talk with the Tyro; The Shooting Positions; Aiming; Trigger Control; Rapid Fire; Wind; Mirage and Light; The Question of Equipment; Sighting our Rifle; "Commence Firing"; Long Range Rifle Shooting; Free Rifle Shooting; Pan American Rifle Shooting; Sporting Rifle Shooting; Estimating Distance; Physical Handicaps; The Rifleman's Eyes; Coaching and the Rifle Team; Targets and Ranges; Organized Rifle Shooting; The Rifle on African Game; The Rifleman in Battle; plus an appendix on rifle coaching.

Frontis. of Samuel Woodfill, decorated WWI officer sighting an M1903 rifle at Camp Perry. Illustrated with 100 photographs, tables and charts.

SAMWORTH PROMOTIONAL: "The most modern and complete work in existence relative to the shooting of high-power military and hunting rifles. This book tells you how to shoot; it is devoted to nothing else but shooting the rifle and it certainly goes deeply into its topic. Here is all about the use of long range military rifles over the standard courses of fire, all that is known of target shooting with the intricate and precise match and 'Free' rifles and all that you will need to know to make you a fast and deadly shot with the modern high-powered hunting rifle.

"A vast amount of original and unknown data is contained between the covers of this book. Not only is the rifle shooting part taken care of thoroughly but all the allied essentials are gone into. Take for instance the question of 'The Rifleman's Eye'; Crossman has included an entire chapter on the optical principles of eyesight and its requirements as applied to shooting. You can find out from this chapter just why your eyes blur when you look through the sights and you can also take the chapter to your local oculist and teach him enough about your shooting requirements that he is enabled to correctly prescribe for your optical ailments. And there are 21 other chapters in the book fully as original and as valuable to you. Take a chapter 'Physical Handicaps'; many enthusiastic shooters must shoot from the left shoulder or have but one arm or are getting along in years; this chapter will prove a revelation to such. Of course there are the usual chapters devoted to shooting from the standard and orthodox positions; we merely mention these odd ones to show how original the author has been with this work and how completely he has covered his subject. There is nothing known today about modern rifle shooting which is not included within the covers of this book. 500 pages with 100 illustrations taken especially to go with the text. Bound to match all our standard works."

SUBSEQUENT PRINTINGS AND ISSUES:

c1932 SATPCO Marines, Onslow County, NC. No ad page date. Tall 8vo. Mustard yellow woven silk embossed filled cloth. Gilt lettering on front and spine. Decorative gilt head- and tailstriping. "STP CO." in gilt above tailcap. Front has decorative embossing of a soldier taking aim with a M1903 rifle. Dust jacket illustration in orange ink on white stock of cowboy taking aim from the sitting position with a rifle while his horse stands nearby. Issued 1935.

c1932 SATPCO Marines, Onslow County, NC. No ad page date. Tall 8vo. Mustard yellow woven silk embossed filled cloth. Gilt lettering on front and spine. Decorative gilt head- and tailstriping. "Samworth" in gilt above tailcap is the only characteristic that distinguishes this issue from that above. Front has decorative embossing of a soldier taking aim with a M1903 rifle. Dust jacket illustration in orange ink on white stock of cowboy taking aim from the sitting position with a rifle while his horse stands nearby. Believed to have been issued after 1941.

SUBSEQUENT EDITIONS:

No known later Samworth editions.

Facsimile reprinted in a deluxe quarter-leather-bound limited edition of 1500 copies by Wolfe Publishing Co. in the Wolfe Library Classics series.

NOTES: This is the second Crossman Book published in 1932 by Samworth. As noted in #5, *The Book of the Springfield*, Samworth approached Crossman to pen a book on centerfire military and sporting rifle shooting. The 1903 Springfield was so popular at the time that, in Crossman's own words in the Foreword, "It became obvious that in treating the Springfield and its many cousins of commercial type, and on the matter of military and sporting rifle shooting . . . would become too much of a field for any one book. And the 'book' became twins; the first, *The Book of the Springfield*, the second this volume." In *Military and Sporting Rifle Shooting*, Crossman improved on earlier works by others such as Whelen's *Suggestions to*

Military Riflemen and Dr. Walter Hudson's *Modern Rifle Shooting*, both of which he refers to. While concerned with contemporary bolt action rifles, this book nonetheless is exhaustive in its content and presentation. Crossman was a very capable author; his books are eminently enjoyable reading and this one is no exception.

In criticism, it can be said that Crossman's military and target shooting background shows through in this work as it does in the other two (ref. #4 and #5) and, hence, the hunting content is comparatively short. However, this takes nothing away from the book; what is presented is indeed thorough.

It is interesting to note that the Onslow imprints have two cases, one a dark kelly green and being the first impression, the other a later-issue mustard yellow. A dark kelly green cased Onslow printing is in the compiler's collection with an inscription by Crossman dated 1932. Two different mustard yellow cases have been seen, the latest having "Samworth" in gilt at the base of the spine.

SAMWORTH BOOKS

7) The American Wild Turkey
James A. Rogers Library collection

7) DAVIS, Henry Edwards

THE AMERICAN WILD TURKEY

c1949 Thomas G. Samworth Georgetown, SC. 1949 title page date. 1948 ad page date. Royal 8vo.

CASE: Unfilled golden-straw-colored coarse buckram. Gilt lettering overprinted on dark green title panels on front and spine. "Samworth" in gilt above bottom of spine. Dust jacket has color illustration of turkey in gliding flight past a moss-draped tree.

CONTENTS: 328 pp. with ads. Twenty chapters incl: Qualifying the Witness; History of the Turkey; What is a Wild Turkey?; Past and Present Distribution; Characteristics; Habits, Food and Cover; Roosting and Feeding Habits; Breeding and Nesting; Boyhood Adventures; Methods and Hunting; Calls and Callers; Outwitting the Hunter; Hermit Gobblers; Shotguns for Turkeys; Rifles for Turkeys; Telescopic Sight for Turkey Rifles; Shotgun and Rifle Contrasted; Hits and Misses; Destructive Factors; Survival and Perpetuation.

Color frontis. of dust jacket illustration by W. A. Weber. E. Stanley Smith gravure illustrations. Some B&W photos.

SAMWORTH PROMOTIONAL: "This is a major contribution to the hunting lore of America, with special emphasis on the use of modern firearms when employed on the grandest game bird on earth. For, in whatever section of America the wild turkey be found, there he outranks all other game--bear, deer, waterfowl or birds all take a position in back of Sir Turk--and any sportsman who has had the opportunity to consistently hunt and kill turkeys will give them first place. Even the most dyed-in-the-wool deer hunter will condone upsetting the drive with a shot at a wild gobbler.

"The author of this comprehensive work has reached his allotted three-score-and-ten years, and this entire span of life has been intimately lived in the habitat of the pure-strain wild turkey; in fact the last three-score of those years have been spent mainly, on and off, in the thought, pursuit and shooting of turkeys. Turkey hunting has been the main passion of Henry Davis' existence and he is known far and wide as one of South Carolina's brag turkey hunters, as well as THE outstanding authority in this State on that noble game bird. He was born on a plantation on the Santee River located in one of the finest bits of turkey range in the entire world and was hunting and killing turkeys by the time his folks put him in long pants.

"In this book, Author Davis gives you the results of more than fifty years of continuous study and pursuit of the American wild turkey, with particular attention and much advice as to the rifles, shotguns and ammunition best adapted for his chosen sport. His knowledge of firearms and ballistics is extensive and has been supplemented by his many years of practical application in the hunting field; readers will appreciate his authoritative treatment from these angles. He writes with enormous zest and a wealth of personal anecdote. He is a rifleman of national renown, as well as a gunsmith of exceptional ability, and his chapters on Shotguns for Turkeys, Rifles for Turkeys, and Telescopic Sights for Turkey Rifles will definitely prove of value to the modern hunter of any game.

"In the early chapters, the author gives a complete history of the wild turkey. He compares the bird with the domestic turkey, outlines its past and present distributions, characteristics, food, cover, roosting, feeding, breeding and nesting habits. One chapter deals with turkey hunting boyhood adventures; others with methods of hunting, calls and callers, hermit gobblers and means utilized by turkeys in outwitting hunters.

"The chapter on Turkey Calls and Callers is an outstanding one--full book value in itself to any turkey hunter. Thorough--and fully illustrated--it describes and shows all of the many different types of turkey calls and then tells of their uses, virtues and faults. Included also is much qualified instruction on how to make your own calls. The art (and it is an art) of calling is fully gone into and Davis tells you just how to call, when to call, how seldom to call and, above all, when not to call.

"Although mainly a hunting book, this work treats in a thoroughly practical way the matter of perpetuating and increasing the species. If its suggestions and recommendations in this regard are put into effect, the wild turkey not only can be restored to its former abundance in many of the coverts of its range from which it has disappeared, but can be made a major game asset in most of the states to which it is native. As Herbert Ravenel Sass, widely known naturalist and writer of the South Carolina Low Country, writes in his preface to this work, '. . . the book will continue to be for many years not only a delight to all who love the sports of the woods but also a standard source of knowledge in its field.' It is a deluxe book--319 pages of text with nine special gravure plates of the wild turkey drawn by E. Stanley Smith."

SUBSEQUENT ISSUE:

c1949 Thomas G. Samworth Georgetown, SC. 1952 Ad page date. Royal 8vo. Later issue. Case and dust jacket identical to that of the first impression above.

SUBSEQUENT EDITIONS:

No Later Samworth editions are known to exist. The title has been exactly reprinted by Stackpole and again in 1984 by Old Masters Publishers of Medon, TN. The latter is a quality photo-offset reprint that has a binding style and color scheme faithful to the Samworth edition; remarkably the name "Samworth" appears in gilt above the tailcap.

NOTES: Herbert Ravenel Sass' words in the preface have proven to be as prophetic as if spoken by Isaiah. This book is yet today recognized as the definitive work on the wild turkey, and is much sought after by hunter, conservationist and collector. It is perhaps the most famous of the Samworth titles. Many sporting bibliophiles know the text well but do not recognize the name Samworth. Despite two Samworth issues and the Stackpole and Old Masters editions, this title is extremely scarce due to the huge demand. Once acquired, the Davis is almost never disposed of; many collectors invest in duplicates to minimize wear on their "keepers."

H. Paul Dove, Jr. Director of the James A. Rogers Library at Francis Marion College in Florence, SC, has had the pleasure of many conversations with Mr. Sam and relates this story: "Samworth had originally contracted with another prominent South Carolina sportsman and author to write a book on the wild turkey, but was extremely dissatisfied by the result, which he called 'thrice-told turkey tales'. A neighbor and friend of Samworth's, Alex Quattlebaum, introduced him to a Florence, SC lawyer, Henry Edwards Davis, who agreed to write a manuscript. Samworth was so pleased with the finished work that he doubled the $1000 fee."

Little more can be said about the book; its high demand has commanded matching prices for years, for both Samworth and Stackpole editions alike. It is the highest-priced Samworth title on the used market.

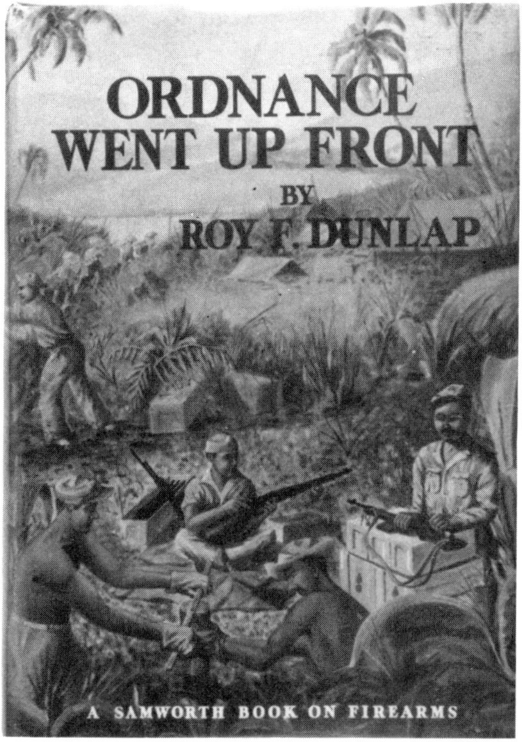

B. R. Smith collection

8) DUNLAP, Roy F.

ORDNANCE WENT UP FRONT Some observations and experiences of a sergeant of ordnance, who served throughout World War II with the United States Army in Egypt, the Philippines and Japan--including Way Stations. With comment and opinions on the many different small-arms in use by the forces engaged.

c1948 Thomas G. Samworth Plantersville, SC. 1948 ad page date. Crown 8vo.

CASE: Medium-gray lightly filled coarse buckram. Gilt lettering on front and spine. Decorative gilt head- and tailstriping. "Samworth" in gilt above tail-cap. Dust jacket illustrated with Ned Smith color painting of soldiers cleaning and repairing weapons in jungle bivouac.

CONTENTS: 418 pp. including index, with several ad pages. Forty-seven chapters incl: "Approach"; Signed Up; On the Move; To Egypt Via South Pole; Desert Station; Modern Egyptian Life in Cairo; With the Eighth Army; At the Captured Equipment Depot; British and German Ammunition; Italian Small Arms Ammunition; Italian Machine Guns; German Small Arms Used in Egypt; The German Light Machine Guns; The German Pistols; "Foreign" German Pistols; The European Pocket Pistols; The Italian Pistols; British Small Arms; A Detail With QM; Rifle Practice in Egypt; With the Rifle Range Detail; Our International Rifle Matches; Stateside Bound; Ordnance School and Mississippi; To the Philippines and the 1st Cavalry Division; Beachhead Landing on Leyte; Early Operations on Leyte; "Fightin' in Them Hills"; The Filipinos; An Ordnance Dump in the Hills; Out of the Hills; Field Repairs and Replacements; The Landing and Fighting on Luzon; Mopping-up on Luzon; Loading Up for Japan; Japan; Japanese Small Arms; The 7.7mm Jap Rifle; The Jap

Pistols; Jap Machine Guns; U.S. Weapons; Miscellaneous European Small Arms; Miscellaneous Musings on Men, Officers, War; Reconversion; The Strength and Suitability of Military Rifle Actions; German Small Arms Manufacturing Plants and Codes; Metric-Inch Conversion Data. Index.

Frontis. of American, German and Japanese bolt-action rifles. Illustrated with 44 plates.

SAMWORTH PROMOTIONAL: "In reality, this book is the elaborated technical diary of an American gun nut who served as an enlisted man in the 27th Ordnance Company of the United States Army and whose duties consisted mainly of going over practically all types of the rifles, pistols and machine guns of World War II, both enemy and allied. The author's services in both Eastern and Western theaters of war puts him right up into the 'been everywhere and seen everything' classification. He really did get around a bit; Aberdeen Proving Ground; Egypt; Mississippi; New Guinea; Leyte; Luzon and Tokyo.

"Working on all types of small arms, usually right in the combat area, was Roy Dunlap's job during most of the war, and he consequently describes and treats these various arms within the light and knowledge of his own first-hand experience--plus that of the men up on the firing line.' When a particular weapon is described, it is actual performance you are reading about--those details of suitability, report, recoil, feel and actual results, known only to practical shooters--and the type of knowledge so desired by other shooting men. These are the facts YOU want to know.

"To the owners of these foreign military weapons--the veterans who brought them back as trophies or the individuals who have since purchased them--*Ordnance Went Up Front* is the book to tell you everything about that weapon; its specifications; the cartridge it shoots; its limitations; and in particular, what can be done with it to make it up into a top-grade sporting rifle or target arm.

"Herein are the most detailed and informative descriptions of German, Italian, British, American and Japanese small arms that have so far been published, with much additional data given on the ordnance of many of the smaller nations. Weapons and cartridges are treated extensively, and in their proper relation to each other. This is the book many shooters thought they were getting when they purchased earlier works on these subjects. It is the first such work really written from the practical-user and not the catalog-reader standpoint, with Dunlap's previous training as an experienced gunsmith and Camp Perry marksman standing him in good stead. He knows what you want to know--and WRITES it. Here is a trained rifleman's qualified opinions, formed after having examined, dissected, repaired and fired all types of modern military weapons used in the past war."

SUBSEQUENT PRINTINGS AND ISSUES:

c1948 Thomas G. Samworth Georgetown, SC. 1953 ad page date. Short crown 8vo. Later issue. Case is a medium-gray slightly filled coarse buckram. Gilt lettering on front and spine. Decorative gilt head- and tailstriping. "Samworth" in gilt above tailcap. Dust jacket illustrated with E. Stanley Smith painted scene of soldiers cleaning and repairing weapons in jungle bivouac. Case and dust jacket are identical to that of the first impression.

SUBSEQUENT EDITIONS:

NOTES: One of the rarest Samworth books, this title served as a text for post-war armed forces ordnance officers and trainees. A large part of the few copies seen on the used and rare book market are ex-library from one military installation or another. It is a work of particular interest to the student of military small arms. The descriptions of the numerous Allied and Axis weapons are excellent. The author's personal experiences are intricately combined with technical facts and objective and subjective observations regarding military life, combat, weapons, etc. This is one of the few works of a "personal experience" nature that documents Axis ordnance from both theaters. Extensive commentary is made on the weaponry of most of the allied nations as well.

Dunlap's *Ordnance* complements and slightly mirrors the other three "war" Samworths; see #12, *Shots Fired in Anger*; #28, *A Rifleman Went to War*; and #35, *With British Snipers to the Reich.* On the used book market it is seldom seen in the dust jacket.

Ordnance Went Up Front

SAMWORTH BOOKS

9) Gunsmithing
M. L. Biscotti collection

9) DUNLAP, Roy F.

GUNSMITHING A manual of firearms design, construction, alteration and remodeling. For amateur and professional gunsmiths and users of modern firearms.

c1950 Thomas G. Samworth Georgetown, SC. 1950 title page date. 1950 ad page date. Crown 8vo.

CASE: Mint-green unfilled buckram. Dark green lettering on front and spine. Decorative green ink head- and tailstriping. "Samworth" above tailcap. Dust jacket illustrated with large drawing of rifle stock diagonally across, with overlay painting illustration on front of late master gunstocker Alvin Linden at work in his shop. "A Samworth Book on Firearms" across bottom of front panel beneath illustration.

CONTENTS: 714 pp. plus ads. Thirty-six chapters incl: The Workshop; Basic Tools; Extra and Special Tools; Reference Books and Sources of Supply; Shop Practice; Helpful Gunsmithing Knowledge; Soldering and Brazing; Welding; Heat Treatment of Metals; Making and Fitting Sights and Accessories; Making, Fitting and Heat Treatment of Parts; Cartridge Information for the Gunsmith; Rifle Barrels and Barrel Information; Chamber and Barrel Work; Rifle Action Work; Pistol and Revolver Work; Shotguns and Shotgun Work; Twenty-two Rim Fire Arms; Browning, Blueing and Blacking of Metal; Fitting Commercial Metallic Sights; The Mounting of Telescopic Sights; Wood for Gunstocks; Design of Gunstocks; Stockmaking-Layout and Inletting; Stockmaking-Shaping and Fitting; Stockmaking-Finishing; Stockmaking-Checkering; Finishing Semi-Inletted Stocks; Stock Repair and Alteration; Styling the Custom Rifle; Custom Metal Work; Ornamentation of Wood and Metal; Target Rifles; The Garand Rifle; Testing Facilities and Apparatus; Tables of Mechanical Reference, Index.

Frontis. of painting of the late Alvin Linden working in his shop. The same illustration is on the dust jacket of both Samworth printings. Text illustrated with plates and sketches.

SAMWORTH PROMOTIONAL: "A 1950 work on gunsmithing--the most complete ever written, containing information on every phase of gunwork from selection of the stock blank on through to the metal engraving and blueing. It is thorough--with not only the 'what' and 'how' but also the 'why'. This work was instigated by the publisher and written at his request by Roy Dunlap with a view of replacing Baker's *Modern Gunsmithing*, which was written back in 1927. Everyone who has read Dunlap's manuscript says it is 'better 'n Baker'.

"The average user of firearms will find this book to be of value in its technical information on barrels, action bedding, accuracy adjustments and trouble corrections. The shotgun information is exceptionally complete and practical: One can read exactly how the job he wants should be done and how to have guns fitted properly to himself so that he can get the most out of them.

"The targetshooter will be interested in the chapters on modern target rifles, their barreling and chamber work, their special furniture and fittings. Dunlap is a hard-boiled, competent rifleman and his information on rifles of precision is backed by trial, experience, quite a few medals and trophies, and many a head of big game.

"The general gunsmith will gain information he has never been able to find except through trial and error. Individual shotguns, rifles, revolvers, and pistols are covered in detail, their weak points mentioned, and instructions given as to how to fix them without the use of a fully-equipped shop and special, expensive machinery. For the first time, complete and official cartridge and chamber specification drawings are published, with headspace data and barrel threadings, on modern cartridges from the .22 long rifle to the .375, including the more popular wildcats. All barrel shank and thread data is shown by drawings as well as dimensions. Barrel specifications and rifling information in all calibers is listed and analyzed.

"Above all--although complete and thorough--this book is not written over the shooter's head. Written and published with the definite aim of turning out the most possible up-to- date information and instruction under one cover--*Gunsmithing* is the best one-book buy that can be obtained today. Its pages are crammed with instruction and formulae necessary in all phases of the gunsmithing art. Professional, amateur, or just plain shooter--this Dunlap work is by far the best shooting-buy offered at the start of the half-century. It has been four years in the making and will prove a milestone in gun literature equal to Baker's famous work of three decades back.

"Gunsmithing is sold under the guarantee that it is better and more applicable today than any other published work on that craft. 800 pages--200 illustrations--36 chapters of the most modern, most complete, best all-around book on gunwork published."

SUBSEQUENT PRINTINGS AND ISSUES:

c1950 Thomas G. Samworth Georgetown, SC. 1952 ad page date. Short crown 8vo. Later issue. Case and dust jacket are identical to those of the first impression.

c1950 Thomas G. Samworth Georgetown, SC. 1955 ad page date. Short crown 8vo. Later issue. Case and dust jacket are identical to those of the first impression. This was the last Samworth issue.

SUBSEQUENT EDITIONS:

This title has been repeatedly reprinted by Stackpole Publishing Co; copies of which are so marked. These issues are usually seen with either mint or medium-dark green cloth cases. Second edition copyright 1964 by Thomas G. Samworth.

NOTES: Dunlap calls this a successor to Baker's *Modern Gunsmithing* in the introduction, and it is a valid description. It makes a very good update, but the amateur or professional gunsmith would want to have both.

A Samworth impression of *Gunsmithing* is hard to find with dust jacket in good condition, since it is a title which would be "used" moreso than read and put on the shelf. Therefore, many copies seen will be soiled as well. Used Stackpole impressions are common; the title is still in print and considered a standard reference.

Gunsmithing

10) Woodchuck Hunter
B. R. Smith collection

10) ESTEY, Paul C.

THE WOODCHUCK HUNTER

c1936 T. G. Samworth Onslow County, NC. 1936 ad page date. 12mo.

CASE: Navy blue thin pebble grain embossed cloth boards. Gilt lettering on front and spine. All edges dark blue ink-washed. Dust jacket is yellow paper with green and black ink depicting a prone "woodchuck hunter" taking aim from a fence line at a small black oblong shape across a field. Jacket flaps advertise two other titles on each, with a list of titles appearing on the back panel.

CONTENTS: 135 pp. Six chapters incl: The Woodchuck; Equipment for 'Chuck Hunting; The Rifle; Telescope Sights; Targeting the Rifle; Ramblings Afield.

No frontispiece. Illustrated with a few B&W plates, and sketches by Philip Plaistridge.

SAMWORTH PROMOTIONAL: "A most entertainingly written and applicable volume for the thousands of 'chuck hunters. All about the woodchuck--how he lives--and how to connect with him at long range with modern rifles. Stalking dope also. Complete data on such new 'chuck cartridges as the .220 Swift--.22/3000--.22 Niedner Magnum and every other worthwhile load. A mass of reloading data--hunting trajectories--accuracy figures for the gun cranks. Any rifleman needs and will enjoy this 136 page well illustrated volume."

SUBSEQUENT PRINTINGS AND ISSUES:

c1936 T. G. Samworth Onslow County, NC. 1936 ad page date. 12mo. Case made of royal blue thin pebble grain embossed cloth boards. Gilt lettering on front and spine. All edges dark blue ink-washed.

Dust jacket is yellow paper with green and black ink and illustrated like that of the early issue described above. Jacket flaps have ads for four other titles on each flap, with a list of titles appearing on the back panel.

c1936 T. G. Samworth Onslow County, NC. 1936 ad page date. 12mo. Case is a smooth, royal blue fine-weave cloth. Gilt lettering on front and spine. All edges dark blue ink-washed. Dust jacket is yellow paper with green and black ink depicting a prone "woodchuck hunter" taking aim from a fenceline at a small black oblong shape across a field. Jacket flaps have ads for four other titles on each flap, with a list of titles appearing on the back panel.

NOTES: *Woodchuck Hunter* is one of the "small Sams" brought out for $1.50 and up in an effort to sell more books in the post-Depression era. The earlier Marshallton and Onslow works sold for $4.00 to $7.50, which was more than a week's wages for some people. *Woodchuck Hunter, Handloader's Manual, Sixguns and Bullseyes, Big Game Rifles and Cartridges, Telescopic Rifle Sights* and a few other titles filled out the "low end" of the Samworth catalog and made quality firearms literature affordable to the average shooter.

Woodchuck Hunter enjoys the honor of being the first book ever published concerned solely with hunting the animal. As claimed in the promotional, it does contain a wealth of information, written by a man who completely understood his subject. It is an easy-reading text, and one that students of sniping would also find interesting. The author was a close friend and shooting partner of Col. Townsend Whelen, who appears with the author in one of the illustrations.

Because of the content, owners of this book seldom part with it once read; this is one explanation for its relative scarcity. It was a fast and popular seller for Samworth; 5500 copies were ordered in two printings and were distributed in the three issues.

The Woodchuck Hunter

SAMWORTH BOOKS

11) Elementary Gunsmithing
B. R. Smith collection

11) FRAZER, Perry D.

ELEMENTARY GUNSMITHING
A manual of Instruction for Amateurs in the Alteration and Repair of Firearms.

c1938 Thomas G. Samworth Onslow County, NC. 1938 ad page date. 12mo.

CASE: Medium-brown pebble grain embossed cloth. Gilt lettering on front and spine, often seen having turned a dull medium to dark green. All edges brown ink-washed. Dust jacket illustrated with a gunsmith at work as viewed through a brick front shop window carrying the lettering "Elementary Gunsmithing".

CONTENTS: 208 pp. plus ads. Foreword plus seventeen chapters incl: Starting in Businesslike Fashion; Models, Textbooks and Reference Works; The Shop and Its Equipment; Necessary Tools and Others; Labor-saving Devices and Gadgets; Materials and Supplies; Routine Jobs; Sights and Scope Blocks; Hunting Telescopes and Mounts; Blueing Barrels and Parts; Soldering and Brazing Metal Parts; Rifle-Stock Jewelry; Restocking Problems; Making Stocks from Solid Blanks; "Warmed Over" Rifle Work; Of Various Matters; Sources of Supply.
Frontis. of a custom .30-'06 Springfield. Illustrated with several B&W plates, sketches and drawings.

SAMWORTH PROMOTIONAL: "Who is better qualified to guide the beginner into the intracacies of guncraftsmanship than a professional gunsmith with more than 50 years practical experience at the workbench? Yep, we think so too, and Perry D. Frazer turned out a masterful work on this most difficult subject.
"Elementary Gunsmithing--a fitting title for a concise, complete manual on amateur gunsmithing--is the ideal 'first book' buy, not only for youngsters, but for

any person just becoming interested and enthused in firearms. Whether this interest is fired by the individual taking up the love of firearms as a hobby or whether he wishes to go a step farther and make it a part or full time business he can get a solid foundation of the fundamentals of gunsmithing in this volume.

"Fitting-out of the necessary workshop and equipment; making special tools; proper methods of tool use; different phases of gunwork, repairs and alterations is all set forth--based upon the actual experiences of the author--along with discussion of special formulae, working kinks, gadgets, et cetera. Illustrations show tools, templates, jigs and necessary fixtures you can make yourself.

"This book thoroughly covers the field of gun remodeling, primary repair work and alterations. Frazer gives detailed instructions on the proper use of essential tools and special fixtures on such jobs as the beginner is qualified to take up. The more common and simple repairs and alterations are discussed and then, in turn, stockmaking, checkering and finishing. Considerable space is devoted to the remodeling of military rifles into modern hunting models--everything from the substitution of a front sight to the turning out of a sporter of the latest type. It takes the workman from buttplate to muzzle and gives a thorough basic knowledge in gunsmithing practice.

"Elementary Gunsmithing is truly the beginner's bible. 210 pages profusely illustrated with full page sketches and many composite halftone plates."

SUBSEQUENT PRINTINGS AND ISSUES:

c1938 Thomas G. Samworth Plantersville, SC. 1944 ad page date. 12mo. Later impression. Case is medium-brown fine-weave smooth cloth. Gilt lettering is often seen turning green. Text paper is gray-white and uncalendered.

Dust jacket illustration is identical to that of the first impression, but has ads for different titles.

c1938 Thomas G. Samworth Plantersville, SC. 1946 ad page date. 12mo. Later issue. Case is a navy blue-black woven silk cloth. Gilt lettering on front and spine. Text appears to be printed on the same type of stock as that above. Dust jacket illustration is identical to that of the earlier impressions, but ads are different; titles printed post-1938 and -1944 are described.

SUBSEQUENT EDITIONS:

A Stackpole reprint has been examined, so marked.

NOTES: This work was commissioned by Samworth to fill the vacancy in his product line for a "low end" inexpensive book on gunsmithing for the beginner. The only other Samworth gunsmithing text at the time was Baker's *Modern Gunsmithing*, #3, a comparatively expensive $4.00. When published in 1938, the Frazer book fit the bill nicely at $2.00. Both Samworth and the author intended the text to be written for the amateur who would be without machine tools, and this book is a success in that regard. One would ordinarily not expect too much of such a work. However, *Elementary Gunsmithing* does have its share of useful tips and information.

The first printing has the imitation pebble grain embossed leatherette cover which is prone to cracking and flaking off in the hinge area. Many specimens display this wear to some degree, with the cloth showing through.

Considered to be a "working" title, it is more often than not seen without a dust jacket. The dust jacket art poses an interesting question. The illustration depicts a gunsmith at work over a rifle in a vise, as viewed through a storefront window on which is lettered "Elementary Gunsmithing", the title. One cannot help but wonder, who would feel comfortable taking their firearms to a business with a name like that?

12) Shots Fired in Anger
Richard J. Smith collection

12) GEORGE, Lt. Col. John B.

SHOTS FIRED IN ANGER

A Rifleman's-Eye View of the activities on the Island of Guadalcanal, in the Solomons, during the elimination of the Japanese Forces there by the American Army under General Patch whose troops included the 132nd Infantry of the Illinois National Guard, a combat unit of the Americal Division, in which organization the author served while encountering the experiences described herein.

c1947 Thomas G. Samworth Plantersville, SC. 1947 ad page date. Crown 8 vo.

CASE: Scarlet-red filled buckram, gilt lettering on front and spine, Decorative gilt head- and tailstripping. "Samworth" above tailcap. Dust jacket illustration by Tsurita depicting Japanese soldiers in "banzai" attack with a U.S. soldier returning fire from a foxhole in foregound.

CONTENTS: 421 pp. plus ads. Foreword and thirty-nine chapters in four parts incl: A Hobby; Fort Sheridan, July '38; Camp Perry, September '39; Training Days; Shoving Off for the Pacific; Beachhead Existence; Henderson Field; Sniping in the Point Cruiz Groves; Night on Guadalcanal; Fighting Beyond the Beachhead; Hurry Up and Wait; Raisin Jack; Approach to Battle; The Fight for Mount Austen; A Shore to Shore Operation; Mop Up on the Tanamboga River; Rest for the Weary; Bushwackin'; Back to the Beachhead; Patrol to the Boondocks; Again?; Out of it; Japanese Use of Rifles; Japanese Rifles; The Model 1939 7.7mm Rifles; The "Thirty Year" Carbine; The Model 44 6.5mm Carbine; The M38 6.5mm Sniper Rifle; The Nambu Pistol; The "Baby Nambu"; The Nambu Lights; The Original Nambu; Japanese Grenades; Japanese Mortars; Japanese Artillery; Anti-Tank Guns; Japanese Small Arms Ammunition; Our Own Guns; The Next Time.

Frontis. of dust jacket illustration by Tsurita. Illustrated with sketches by Hideo.

SAMWORTH PROMOTIONAL: "Johnny George is remembered by Illinois riflemen as a State Team member at Camp Perry; as the youngest .30 caliber champion the State ever had; and as a many- times-winner in various local and regional rifle competitions. He knew plenty about the peace-time use of rifles. He got into World War II early, as a lieutenant in the 132nd Infantry, Americal Division, on Guadalcanal. This was the first Infantry Division to be used aggressively against the Axis Powers in any theatre. After the Guadalcanal campaign, he volunteered and went through Burma with "Merrill's Marauders." Subsequent wartime assignments took him more than twice around the globe. He saw PLENTY of shooting.

"*Shots Fired in Anger* is a book covering George's rifle shooting experiences up to and including his platoon leader and sniping ventures on Guadalcanal. He wrote it particularly for other riflemen. Half of the text of this work is in narrative form, giving a shot-by-shot account of Lt. George's part in the Battle of Guadalcanal and the events leading up to it. He discusses rifleman training and tactics first, then tells the actual stories of how he and his fellow Infantrymen killed the enemy with rifles and other Infantry weapons. Throughout the book George disclaims being a warrior, but he seems to remain somewhat this side of those who condemn War as a monotonous bore, devoid of thrills. His account sometimes takes the slant of a good big game hunting story as told by one alive to the dangers, as well as the thrills, of the chase. One thing he makes clear is the fierce pride he takes in having been a doughboy and a rifleman. He believes ardently in the traditions of the foot soldier--and more than ever he believes in accurate rifle fire.

"The second half of this book, entitled 'The Tools Used', gives a gun-nut's description of the weapons used and encountered in the field, relating their manners of performance. This presentation almost brings the Japanese and American weapons to life; George shows here that he certainly regards guns as much more than inanimate bits of metal and wood.

"Every American rifleman and member of the National Rifle Association should read this book. Aside from its entertainment value, it contains a world of practical information which will prove of worth to the younger members of our shooting clan, who may well be called upon to use our American military rifles in another war. It tells you just what to expect of the modern military rifle when shooting 'for keeps'--and also what to look out for from the rifle of the other fellow. The sketches illustrating this work were done by one of the best artists in Japan."

SUBSEQUENT PRINTINGS AND ISSUES:

A second Samworth printing is believed to exist, but it is unconfirmed at present. All copies of this title examined to date have been identical to the first impression described above.

SUBSEQUENT EDITIONS:

Currently available in an updated and expanded edition from the National Rifle Association, now in later printing, and contains much additional material on the author's exploits in Burma with Frank Merrill's "Marauders".

NOTES: This title is the second of the four "war" Samworths, the others being #8, *Ordnance Went Up Front* by Dunlap; #28, *A Rifleman Went to War* by McBride; and #35, *With British Snipers to the Reich* by Shore. Like the others, it is a first-person account of one's experiences in war. *Shots Fired in Anger* is particularly interesting in that it chronicles the exploits of an officer who was already an accomplished rifle shot. The text references to the author's sniping experiences are noteworthy. The fourth part of the book gives a detailed description of Japanese arms with some observations on American weapons, and makes an excellent accompaniment to those of Dunlap in *Ordnance Went Up Front.*

Shots Fired in Anger is one of the scarcer Samworths. Book dealers specializing in hunting and firearms titles sometimes list it at lower prices than those dealers who trade in the military history genre.

The gilt lettering on the maroon buckram is almost universally seen to tarnish. Copies with bright lettering are scarce. The dust jacket is printed on a comparatively light basis weight paper stock, and thus they usually exhibit a higher degree of wear, chipping, and tearing.

SAMWORTH BOOKS

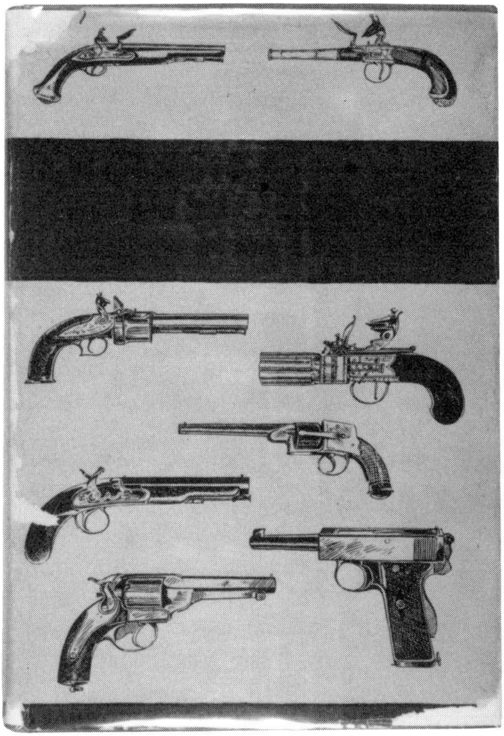

13) English Pistols and Revolvers
B. R. Smith collection

13) GEORGE, John Nigel

ENGLISH PISTOLS AND REVOLVERS

An Historical Outline of the Development and Design of English Hand Firearms from the Seventeenth Century to the Present Day.

c1938 Thomas G. Samworth Onslow Co, NC. 1938 title page date. 1938 ad page date. Royal 8vo.

CASE: Gray filled woven silk cloth; gilt lettering overprinted on red title panels. Gilt "Samworth" in red panel at bottom of spine. Dust jacket has horizontal title band and sketches of several early and late English handguns. Pages are untrimmed.

CONTENTS: 256 pp. plus ads. Foreword plus twelve chapters incl: English Pistols and the Civil Wars; The Commonwealth, Restoration and Revolution; First Half of the Eighteenth Century; Last Half of the Eighteenth Century; The Last of the Flint-Lock; Forsyth and the Detonating System; Percussion Pistols; Early Revolving Pistols; Early Percussion Revolvers; Colt and His Imitators; English Percussion Revolvers; The Breech Loading Systems.

Frontis. of two views of Snaphuance Revolving Pistol circa 1650. Illustrated with several plates and some sketches.

SAMWORTH PROMOTIONAL: "Practical shooting men, as well as collectors, will enthuse over this outstanding book. Starting with the distinctive 'English lock' pistol of the time of Charles I, it carries the reader through each successive stage of hand firearm development, usage and improvement. All about the first firelocks, 'scrued pistolls' (sic), horse (holster) pistols, single, double, four-barrel and early repeating flint-locks, Queen Anne pistols, screw-barrel pistols, pocket, belt, carriage, duelling and military models. Next comes the development of the detonating system, with its percussion pistols, pepperboxes and revolvers then made possible. Finally the introduction of the metallic cartridge and perfection of the revolvers and self-loaders of the present day.

"The book is replete with interesting and authentic notes relative to Joe Manton, Durs Egg, Nock, Mortimer and other leading gunmakers of the reigns of the Georges, as well as brief sketches of Elisha Collier, Col. Sam Colt and other Americans who stamped their ideas and influence upon the manufacturing practices and inventions of their times.

"*English Pistols and Revolvers* describes and illustrates each distinct type and development of English hand firearms. The book consists of 256 pages of reading text; there are 201 different pistols and revolvers shown in the 26 full-page plates which illustrate the work. Not the least attractive and valuable features are the many small sketches of period decorations, butt masks, hammers and proof marks which are scattered throughout this book. It is an authoratative work which will remain the standard of reference for collectors, arms students and shooters for years to come."

SUBSEQUENT PRINTINGS AND ISSUES:

c1938 Thomas G. Samworth Plantersville, SC. 1938 ad page date. 1938 title page date. 8vo. Later issue. Gray case is similar to that of the first impression but a darker medium-gray; cloth weave is slightly coarser. Dust jacket illustration identical to that of the first impression.

c1938 Thomas G. Samworth Plantersville, SC. 1938 ad page date. 1938 title page date. 8vo. Later issue. Case is a dark, almost charcoal gray, otherwise identical to that above. Dust jacket illustration is identical to that of the first impression.

SUBSEQUENT EDITIONS:

Additional reprints were made in 1961 by Holland Press and 1962 by Arco.

NOTES: Long recognized and in demand as a standard work on the subject, this volume has high intrinsic value because of its thorough coverage of the topic. John Nigel George, who was killed by a sniper's bullet in North Africa in 1942 at age 39, was a recognized expert on the history of English firearms.

A companion volume, *English Guns and Rifles*, was published posthumously in 1947 by Samworth, #14 which see.

This title is moderately scarce; being frequently listed in dealers' catalogs on the rare market. First impressions in dust jacket are understandably less prevalent. *English Pistols and Revolvers* has the distinction of being the first book Samworth produced in the "Deluxe Sam" format with contrasting title panels letterpress-overprinted in gold.

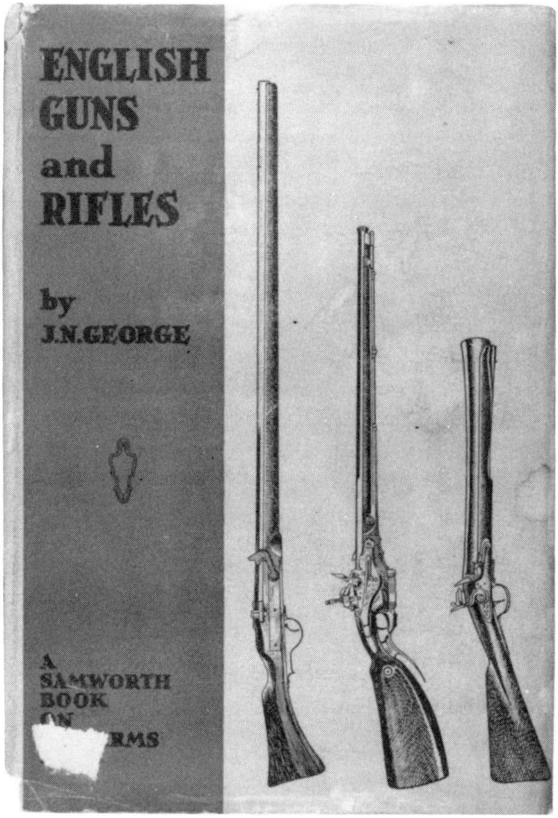

14) English Guns and Rifles
James A. Rogers Library collection

14) GEORGE, John Nigel

ENGLISH GUNS AND RIFLES

Being an Account of the Development, Design and Usage of English Sporting Rifles and Shotguns--From Their Introduction during the Fifteenth Century until the Advent of the Metallic Cartridges in the Nineteenth Century--together with Notes on Muskets, Service Rifles, Blunderbusses and Other Arms including also Various Historical Notes and Accounts regarding Individual Makers and Users of these Arms.

c1947 Thomas G. Samworth Plantersville, SC. 1947 title page date. 1947 ad page date. Royal 8vo.

CASE: Straw-yellow lightly-filled coarse buckram; gilt lettering overprinted on green panels. Gilt "Samworth" in green panel at bottom of spine. Dust jacket has a vertical title bar on the left with an illustration of two rifles and a blunderbuss arranged vertically on the right.

CONTENTS: 344 pp. incl. index. Twelve chapters incl: Introduction of Firearms into Tudor England; Firearms of Stuart England and the Civil Wars; The Commonwealth, Restoration and Revolution; Smooth Bores of the Early Eighteenth Century; The Flint-Lock Rifle: 1700-1780; The Flint-Lock Rifle: 1780-1820; The Sporting Smooth-Bore; The End of the Flint-Lock; The Detonating System; Sporting Rifles: 1830-1860; Service Small Arms and the Coming of the Breech-Loader. Appendix of British gunmakers of the flint-lock period; author's bibliography.

Frontis. of two flint-lock repeating rifles and one breech-loading rifle. Illustrated with B&W plates, and sketches by the author. Has a second frontispiece of the author examining a Colt SAA revolver, with an accompanying brief biographical statement.

SAMWORTH PROMOTIONAL: "This magnificent work traces, in profuse detail and in most entertaining form, the progress of invention and perfection of shoulder firearms from the 'hand gonne' of the 15th century until the coming of the breech-loader of the 1850s. It is the most comprehensive study and analysis of firearms yet produced. Starting with the match-lock caliver of Queen Elizabeth's time, the author depicts progressively the various firearms which followed--the match-lock musket, the wheel-lock, the snaphuance, then the flint-lock in its various forms, and finally the percussion system with its adaptations and uses. Concurrently with each type of weapon is treated the application of such principles as were involved in military arms, fowling pieces, sporting arms and weapons of self defense. Each and every type of firearm is fully presented, with complete and authoritative descriptions and proper illustrations of the development and use of blunderbusses, wall, boat, and turnpike guns, musquetoons, coaching carbines, revolving and repeating rifles of the flint-lock era; breech-loading flint-lock rifles in various forms; screw-barrel and screw-plug systems, 'shot and ball' guns; wildfowl and punt guns; deer stalking rifles; early big-game rifles of tremendous bore for India and Africa; the detonator lock, cap lock, pill locks, pellet and tube locks; then the various degrees by which the percussion gun was perfected. Capping methods and capping breech-loaders, with the various four-barrel, seven-barrel, and other 'pepperbox' modifications are described. Combustible cartridge cases follow, with various applications towards a self contained cartridge loading into the breech of the gun, and the book ends with the coming of the pin-fire cartridge and the breech-loader.

"The work closes with the most extensive listing of English gunmakers so far published, together with their location and date of operation. It is the outstanding work of this day in the history, development, application and use of firearms up to the advent of the breech-loader. 338 pages, 25 plates and 50 technical and period drawings comprise the format."

SUBSEQUENT PRINTINGS AND ISSUES:

No known later Samworth issues.

SUBSEQUENT EDITIONS:

Reprinted using original plates during the late 1950s or early 1960s by Stackpole; copyright 1947, by Thomas G. Samworth. Case bindings are nearly identical to those of the Samworth edition, with the exception of "Stackpole" printed above the spine tail-cap, and a slight variation in shade of off-white buckram. Dust jacket is a mottled mustard color without the rifle illustrations found on the Samworth edition.

NOTES: A companion to #13, *English Pistols and Revolvers*, following largely the same format. Widely recognized as one of the most significant references in the field. The appendix has several pages of early English gunmakers and represents the first such published compilation. The extensive work was published after the author's death; he was killed in action with the British Army in North Africa in 1942.

Extremely scarce title in the Samworth impression, and even moreso with the dust jacket, this book commands high prices on the used market due to steady high demand as an authoratative collector reference. Stackpole reprint editions are more frequently seen than the Samworths, but the pricing seen does not appear to differentiate between the two.

15) Modern American Pistols and Revolvers

B. R. Smith collection

15) GOULD, Arthur Corbin

MODERN AMERICAN PISTOLS AND REVOLVERS An account of the development of pistols and revolvers in America; description of the varieties manufactured; manner of shooting them; work accomplished with these arms; departments of pistol and revolver shooting; impressions formed by studying these arms; and rules governing pistol and revolver competitions.

c1888 by A. C. Gould & Co, and c1894 by Bradlee Whidden.

Republished 1946 by Thos. G. Samworth, Plantersville, SC. 1946 title page date. 1947 ad page date. 8vo. Being a reprint of the 1894 Bradlee Whidden second enlarged edition.

CASE: Forest-green filled woven silk embossed cloth. Title lettering printed on 2 3/4" X 3" paste-on title piece on front cover; illustrated with handlebar-mustachioed man in bowler hat firing a pistol. Other male spectators partially visible in background. Smaller paste-on title piece near top of spine. Pages left untrimmed, with end having ragged "deckle-edge" effect from the paper machine on which the stock was produced.

Dust jacket illustrated with three-color woodcut illustration of the same gentleman, and his spectators are in fill view; several men and women in turn-of-the-century costume standing and seated. Illustration is continued on the rear of jacket showing two other gentlemen at a firing-bench.

CONTENTS: 222 pp. Fifteen chapters incl: Development of American Pistols and Revolvers; American Single-Shot Pistols; The Colt Revolver; American Revolvers--Smith & Wesson's Productions; Miscellaneous Revolvers--Revolvers Classified; Test of Military Revolvers by the United States Ordnance Board; Target Revolvers; Pocket Revolvers; Ammunition for Pistols and Revolvers; Reloading Ammunition for Pistols and Revolvers; Revolver Shooting Record in America; Some Performances with the Pistol; Pistol and Revolver Shooting At Long Range; Impressions Formed by Investigations; Rules for Pistol and Revolver Shooting. Following the preface to the second edition is a six-page biography of Arthur Corbin Gould by Frank J. Kahrs of Remington, written in 1946 for inclusion in this Samworth reprint.

Frontis. of the club house and shooting pavillion of the Massachusetts Rifle Association at Walnut Hill, Woburn, Mass. Illustrated with dozens of woodcuts of pistols, revolvers, shooting positions, ammunition, etc. At the end of the text are reproductions of period advertisements of Winchester, Smith & Wesson, Colt, Lyman, UMC, Stevens and Ideal Manufacturing Co. products and two pages of Samworth ads.

SAMWORTH PROMOTIONAL: "This authoritative and very popular work was first run as a serial in the old magazine 'The Rifle' beginning in the June 1887 issue. It then appeared in book form and, a few years later, was greatly enlarged upon the publication of a second edition. For the next generation, this Gould treatise on American handguns was the standard and only existing work. It is still a most interesting volume, replete with technical and practical notes on all the various makes and models of pistols and revolvers; at that time there were about a dozen different firms making target and hunting handguns. Competitive shooting was one of the leading popular sports and Ira Paine, W. W. Bennett, Annie Oakley and other crack shots were all in the public eye. Gould tells us of the ability and exploits of these great characters of American history who thought little of putting up a stake of $1000, with other bets on the side, to back up their shooting ability in private matches.

"Gould wrote at length on such shooters, their choice of weapons and their shooting style and opinions. Our modern handgun users can study with profit from the many shooting positions illustrated through out this book. Much of his data and shooting information is still applicable and this book was the outstanding work of its day on the handgun and its use. We have reprinted it as it was produced, text and illustration, and the work should be in the library of every arms lover and shooter."

SUBSEQUENT PRINTINGS AND ISSUES:

Four examples of a later issue of this Samworth reprint have been seen with an inexpensive kelly-green fine pebble grain case with black spine lettering and no paste-on title labels. The pages are trimmed and there is no ad page date. The paper stock differs in shade and texture from that of the first printing. The compiler has not seen a dust jacket for these.

NOTES: Back in the days of black powder and lever-action repeating rifles the outstanding firearms writer and authority was Arthur Corbin Gould. It was he who started the original publication called "The Rifle" that evolved into "Shooting and Fishing", then "Arms and the Man" and in the '20s, "The American Rifleman". The second, enlarged edition of his 1888 work on American handguns was faithfully reprinted by Samworth, long an admirer of the accomplished author.

A continuing enigma is the second impression of the Samworth reprint mentioned above. Originally believed to be a one-off library rebinding until three other identical examples were seen. It was reprinted using Samworth's plates, because the data on the title page says "Reprinted 1946 Thomas G. Samworth Plantersville, South Carolina" under the Gould and Whidden copyrights, just like the first impression. Taking into account the book's initial reprint impression was March, 1947 and Samworth's address changed to Georgetown in late 1948, this cheap-looking second issue is certainly either a Georgetown book or a reprint done by another party using the Samworth plates. This is speculation on the part of the compilier, as this second issue has no other publisher's marks. Suffice to say a first impression Samworth reprint can be identified by the dark-green fine cloth covers with the decorative paste-ons, as well as the untrimmed deckle edge text paper.

16) Modern American Rifles
B. R. Smith collection

16) GOULD, Arthur Corbin

MODERN AMERICAN RIFLES With descriptions of processes of manufacturing; appliances used by riflemen for hunting and target shooting; directions for bullet-making and reloading cartridges; positions adopted in various styles of shooting; trajectories of rifles; and a very full record of inventions, improvements, and work accomplished with American rifles.

c1891 Bradlee Whidden

Republished 1946 Thos. G. Samworth Plantersville, SC. 1946 title page date. 1947 ad page date. 8vo.

CASE: Forest-green filled woven silk. Gilt lettering on 2 3/4" X 3" paste-on title piece on front cover illustrated with the heads and shoulders of men viewed from the left rear quarter at a firing line. Smaller paste-on title piece near top of spine. Pages left untrimmed. Dust jacket illustrated with three color woodcut of the same men in full view, with a seated scorekeeper in the foreground. Illustration is continued on the rear of jacket showing another gentleman seated awaiting his turn looking down range at the targets.

CONTENTS: 333 pp. Twenty-four chapters incl: Discovery and Principles of a Rifle--Rifles of the Present--Material for Rifle Barrels; Manufacture of Rifle Barrels--Boring--Straightening--Rifling--Leading and Chambering; Forgings--Additional Work on Barrels--Stocking; Rifle Sights--Front Open Sights; Rifle Sights--Rear Open Sights and Rear Peep Sights; Combination Rifle Sights; Target Sights for Rifles; Telescope Rifle Sights; Hunting Rifles--Single Shot and Repeating Rifles; Target Rifles--Rifles for Off-Hand and Test Target Shooting; Military Rifles; Pocket Rifles; Positions in Rifle Shooting; Aiming, Sighting, Holding and Firing; Trajectories of Rifle Bullets; What it is possible to do with a Rifle; Constructing a Rifle

Range--Laying out a Range--Arrangement of the Butts and Pits--Systems for Marking Shots; Targets used by American Riflemen; Preparing Rifle Ammunition; The Art of Bullet Making; Modern Machinery for Manufacturing Rifles; Round Bullets in Modern Rifles; plus an unnumbered chapter, Rules Governing Rifle Shooting. Following the preface is a six-page biography of Arthur Corbin Gould by Frank J. Kahrs of Remington, written in 1946 for inclusion in the Samworth reprint. Illustrated with woodcuts and pen-and-ink sketches.

SAMWORTH PROMOTIONAL: "This classic of the early '90s has been reprinted by us, word for word, as Arthur C. Gould originally wrote it. It is an outstanding example in early American arms literture and was the rifleman's bible of that decade. This book is our one main record and source of information of the rifles of the black powder days. Full of data, suggestions and advice on the early breech-loading single-shot and repeating rifles and the blackpowder cartridges they fired; also much interesting material about the leading riflemen of those days and the records they established. All the early Winchesters, Ballards, Bullards, Marlins, Stevens, Wurffliens and standard makes of the day are fully described and their merits discussed and compared, with game such as the buffalo and grizzly bear used as the criterion for killing power. These were the days of the paper-patched bullet, the famous Pickett ball for grizzlies and the big bores carrying plenty of soft lead. Their cartridges come in for a thorough going over and the various 'soft coal' loads of those times are analyzed. Shooting positions and methods are fully gone into and illustrated. Handloading and loading tools are included.

"Early range systems are given and information regarding the early days of the National Rifle Association and its matches is all here. It presents a wealth of material relative to the western big-game hunting of the last century and the rifles and ammunition used by leading sportsmen of those days. This is still a useful book and one of great interest to shooters, arms collectors and arms students."

SUBSEQUENT PRINTINGS AND ISSUES:

No known later impressions.

NOTES: This book on American rifles was written by Gould as a companion to his earlier *Modern American Pistols and Revolvers* (#15). Samworth faithfully reprinted both titles which were copyrighted 1946, but they were actually bound in March, 1947. The cases and dust jackets on both titles are well done and match, and they make a handsome pair. This text on rifles is particularly noteworthy because it was perhaps the first comprehensive American book on the subject, published just about the time target shooting came into vogue. For the present-day arms enthusiast with an interest in black powder cartridge rifles, this book contains a bonanza of valuable information. As of this writing, the original Bradlee Whidden imprints are often quoted at three to four times the prices seen for the Samworth edition, the latter representing an outstanding value for those collectors of either period rifles or Samworths.

17) Pistols and Revolvers and Their Use
M. L. Biscotti collection

17) HATCHER, Major Julian S.

PISTOLS AND REVOLVERS & THEIR USE

c1927 SATPCO Marshallton, DE. No ad page. 8vo.

CASE: Maroon silk embossed filled cloth. Gilt lettering "Pistols and Revolvers" on front and spine, gilt decortive head- and tailstriping. Gilt "STP Co." above tailcap. Decorative embossing of a right hand holding a 1911 Colt .45 Automatic on front. Dust jacket illustrated with a B&W halftone photo of a man firing a handgun.

CONTENTS: 399 pp. Eighteen chapters incl: Historical outline; Further Progress--The Development of Modern Types; History of Pistol Shooting; Pistol and Revolver Shooting of Today; Learning to Shoot; Practical Shooting; Hints on Using Hand Arms for Defense; Pistols and Police; General Notes on the Various Arms; Single-Shot and Automatic Target Pistols and Revolvers; Military Revolvers and Automatics; Pocket and Home Defense Arms; Notes on Ammunition; Individual Description of the Most Common Pistol and Revolver Cartridges; Notes on Stopping Power; Reloading Ammunition; Targets, Ranges and Competition; Rules of the U.S.R.A. and N.R.A. Appendix contains nine tables: Average Ballistics of Pistol and Revolver Cartridges as Made in the U.S, 1927; Revolver Cartridges in Order of Penetration; Popular Pistol and Revolver Cartridges in Order of Muzzle Energy; Relative Ballistic Efficiency of Smokeless Pistol and Revolver Cartridges, as Indicated by Loss of Velocity and Energy over 100-yard Range; Cross-Section Area of Bullets in Square Inches; Popular Pistol and Revolver Cartridges Arranged in Order of Estimated "Shock Power" (Sectional Area of Bullet in Square Inches Multiplied by the Muzzle Energy); Estimated Shock Power of Popular Smokeless Pistol and Revolver Cartridges at 100

Yards Compared to That at Muzzle; Status of Cartridges Treated in this Book; American Cartridges for Foreign Automatics.

Frontis. of competitors firing at a section of the pistol range at Camp Perry. Illustrated with B & W halftones, and line-drawing sketches by the author and others.

SAMWORTH PROMOTIONAL; "The standard work of today on the subject of handguns and handgun use is the above book, published in the fall of 1927. It is a work of over 400 pages, illustrated with more than 125 special photographs and line drawings. Every red-blooded American is deeply interested in pistols and pistol shooting and here is a volume which covers the subject completely and accurately. It will appeal to either the skillful target shot or the practical exponent of snap shooting and self defense. Both branches of handgun shooting are fully treated.

"Chapters are included telling all about the various types of single-shot pistols, revolvers and automatic pistols, their virtues and shortcomings and their application to the needs of the shooter. The various models and calibers are fully described. How to learn to shoot, target use and practical shooting, and how to use the weapon in self defense. Special chapters have been written for the police, paymasters, bank clerks and similar operatives whose occupation demands that they familiarize themselves with this means of defense. Every possible phase of pistol and revolver shooting has been properly treated in the extensive and interesting volume. Bound in silk cloth."

SUBSEQUENT PRINTINGS AND ISSUES: None.

Instead of being reprinted, this text served as the basis for a completely new and different volume, *Textbook of Pistols and Revolvers*, published in 1935. See #18.

NOTES: This was Julian Hatcher's second major book, his first being *Machine Guns* published in 1918 by George Banta. *Pistols and Revolvers* is a very interesting work from several different standpoints. The technical content is massive and a must-read for any student of the handgun. Additionally, the book gives an implicit commentary on the politics and mores of 1920s American society. The author has included a graphic account of the famous Thompson-LaGarde experiments that would ellicit interesting responses from the many friends-of-animals groups that exist today. In another chapter, Hatcher advocates the then-legal concealed and loaded carrying of handguns in the event that one is called upon to defend himself or others from attack. Several "case histories" are included in support of his argument. One is compelled to contrast that sort of posture with the present-day policies of our major cities and federal government.

From the shooter's standpoint, all the tips and instruction on how to hold and shoot the handgun outlined in this book are very useful. The author gives exhaustive descriptions of the different makes of pistols and revolvers, as well as the cartridges and calibers chambered in them.

Samworth wrote to Hatcher in late 1930 to give him a "progress report" and a royalty check. He had this to say about the book:

"Good success, a sure seller and a money maker in spite of the fact that it must compete against about five other pistol books. Will run through a steady reprint of editions, provided it is brought up to date with each printing and kept along descriptive and technical lines. Printer short changed me on this book and only delivered 2800 copies."

Thus, the serious collector should be aware that *Pistols and Revolvers & Their Use* is also one of the rarest Samworth titles since it was never reprinted between its initial publication in 1927 and the advent of its "replacement" eight years later.

Last, for the Samworth enthusiast, this book poses problems as do the other two Hatcher titles. *Pistols and Revolvers & Their Use* (1927) is so often confused with *Textbook of Pistols and Revolvers* (1935) it is tragic. The successor is **not** a "second edition"; but a complete rewrite that contains only a bit of the historical information present in the former. Some collectors and handgun shooters have been heard to say that "only the 1935 'edition' is worth having" and this is simply not true. The books differ sufficiently to warrant ownership as a complementary pair.

SAMWORTH BOOKS

18) Textbook of Pistols and Revolvers
N. F. Berringer collection

18) HATCHER, Major Julian S.

TEXTBOOK OF PISTOLS AND REVOLVERS

c1935 SATPCO Marines, Onslow County, NC. No ad page date. 8vo.

CASE: Black silk embossed filled cloth. Gilt lettering on front and spine. Decorative gilt head- and tail-striping. Gilt "STP CO" above tailcap. Decorative cover embossing of crossed 1911A1 .45 Automatic and Colt-like single-action in front of a target. Dust jacket is illustrated with sketches of handgun parts, barrels, sights, etc.

CONTENTS: 533 pp incl. index. Sixteen chapters in three parts: Modern Handguns, General Information; Target Arms and Outdoorsman's Guns; Military Revolvers and Automatics; Pocket and Home Defense Handguns; Technical Notes on Handguns; Accessories; General Information on Ammunition; Interior Ballistics; Exterior Ballistics; Individual Description of Cartridges; Technical Notes on Ammunition; Bullet Effect and Shock Power; Learning to Shoot; Practical Shooting; Shooting in Self Defense; The Peace Officer and His Gun. Marked "Shooter's Edition" on the title page.

Frontis. of the author's son Robert in uniform, firing a S & W .357 Magnum revolver. Illustrated with B&W halftones and line drawings.

SAMWORTH PROMOTIONAL: "Here is the most complete book ever written for handgun users. There are 532 reading pages with some 190 illustrations. Page after page of original technical data is included. The book is divided into three general parts, devoted respectively to hand firearms--their ammunition--their use. Each part is complete and a book in itself. Here are all the essential factors entering into the manufacture and performance of hand firearms; rifling--

barrel length--accuracy--firing pin blow--sights-- trigger pull--lock time--grips--barrel life, etc. Ammunition principles fully treated. Chapters on interior and exterior handgun ballistics. Detailed discussions on recoil--jump--barrel time--chamber pressure--gas action--trajectory. The subject of bullet design and effect, together with its shocking power given particular treatment. A full section devoted to the proper use of both revolver and automatic pistol.

"An essential, wholly accurate and complete work of reference and instruction for any handgun user. Beginner or expert will find it equally interesting and instructive."

SUBSEQUENT PRINTINGS AND ISSUES:

c1935 SATPCO Marines, Onslow County, NC. No ad page date. 8vo. Later issue. Black silk embossed filled cloth. Gilt lettering on front and spine. Decorative gilt head- and tailstriping. Gilt "STP CO" above tailcap. Decorative cover embossing of crossed 1911A1 .45 Automatic and Colt-like single-action in front of a target. Dust jacket is illustrated with sketches of handgun parts, barrels, sights, etc. It is printed in black ink on oyster-white stock, and lists other titles on the back panel. The first issue of the first impression described above has several errata, and this second Onslow issue has an errata sheet bound in usually on the verso of the title page. It is suspected that these errors were discovered after the first printing was completed, perhaps partially.

c1935 SATPCO Marines, Onslow County, NC. No ad page date. 8vo. Later impression. Black woven-silk embossed filled cloth. Gilt lettering on front and spine. Decorative gilt head- and tailstriping. Gilt "STP CO" above tailcap. Decorative cover embossing of crossed 1911A1 .45 Automatic and Colt-like single-action in front of a target. Dust jacket is illustrated with sketches of handgun parts, barrels, sights, etc., and printed in black and gold ink. Other titles advertised on back panel. Examples of this still later impression have the errata corrected in the text itself. The case on this second printing is similar to that of the first described above, but the embossed cloth is more a vertically-oriented, fine-line texture. Additionally, the text of this second impression is printed on lighter basis weight stock, making the book one-fourth thinner than either issue of the first impression.

For those interested, the first issue errata are reprinted below:

ERRATA

Page 165, Delete the word "not" in 16th line from bottom.
Page 166, Firing pin blow of Iver Johnson should be .0085 instead of .085.
Page 170, Line 9, "cartridges" should be "cartridge".
Page 221, Line 6 from bottom, "s" carried over to next word from "sometimes".
Page 228, Tabulation should read 9mm Luger instead of 9mm Mauser.
Page 310, "the" should be inserted between lines 7 and 8.
Page 319, Line 3, end of line insert "of 810 f.s."
Page 327, Line 3, from bottom, no "s" on "kind."
Page 330, Bullet weight of .22 WRF is 45 grains, not .45.
Page 350, Under tabulation of penetration "Drift, Inches" should be "Depth, Inches".
Page 426, End of 2nd par. no quotes.
Page 450, Line 13, "trigger guard" should read "butt".
Page 376, Line 12, group diameter should be "5.59" instead of ".559".

c1935 SATPCO Plantersville, SC. 1944 ad page date. Later impresssion. 8vo. Black woven-silk embossed filled cloth. Gilt lettering on front and spine. Decorative gilt head- and tailstriping. Gilt "STP CO" above tailcap. Decorative cover embossing of crossed 1911A1 .45 Automatic and Colt-like single-action in front of a target. Dust jacket is illustrated with sketches of handgun parts, barrels, sights, etc. like that of the first impression. This printing is nearly one and one-half times as thick as the previous issues, owing to the use of bulky, high-caliper text stock. The case binding is virtually identical to that of the first impression, with the exception of the woven silk binding cloth texture.

SUBSEQUENT EDITIONS:

Bound in one volume with *Textbook of Firearms Investigation, Identification, and Evidence,* first published concurrently in 1935; see #19.

Facsimile reprinted in deluxe quarter-leather-bound limited edition of 1500 copies by Wolfe Publishing Co. in the Wolfe Library Classics series.

NOTES: This is the Hatcher title which has been so sought after during the last forty-odd years, and for good reason. Written by a recognized firearms authority, *Textbook of Pistols and Revolvers* has been cited as a reference in innumerable articles and other books. For anyone wanting to learn about the subject, this is the "Textbook" for Handgun 101.

It has unfortunately overshadowed its predecessor of nearly identical title, *Pistols and Revolvers & Their Use*, #17, an excellent text in its own right. The contents of this 1935 work represent a rewrite of considerable departure from the earlier book.

Further confusion arises from the fact that a different work published concurrently, #19, *Textbook of Firearms Investigation, Identification, and Evidence* has a similar grandiose-sounding title. *Textbook of Pistols and Revolvers* is marked "Shooter's Edition" on the title page to differentiate it from this more forensically-oriented book.

A difficult task presents itself when one attempts to determine if an Onslow County copy is a first issue of the first printing or not. For the uninitiated, the only sure issue is the one with the bound-in errata sheet. Copies with uncorrected errata can be considered first issues in the absence of evidence indicating a missing errata sheet has been removed. In an attempt to make the task of identification easier, the compiler has reprinted the errata above.

A further item of note: Evidently the gilt lettering and the "silk" case material used in the later, thinner Onslow County books are incompatible. More often than not, examples are seen to exhibit worse-than-usual flaking and wear of the spine lettering. Therefore, "FINE" condition specimens of these issues are very rare.

19) Textbook of Firearms Investigation, Identification, and Evidence
B. R. Smith collection

19) HATCHER, Major Julian S.

TEXTBOOK OF FIREARMS INVESTIGATION, IDENTIFICATION AND EVIDENCE together with the *Textbook of Pistols and Revolvers*

c1935 SATPCO Marines, Onslow County, NC. No ad page date. 8vo.

CASE: Black fine weave embossed linen. Gilt lettering on front and spine. Decorative gilt head- and tailstriping. Decorative cover embossing of crossed M1911A1 .45 Automatic and Colt-like single-action in front of a target. Dust jacket illustrated with four bullets and an early comparator microscope. Red printing proclaims foreword to the book by J. Edgar Hoover.

CONTENTS: 342 pp. in the title text, eleven chapters incl: The Firearms Expert and the Court; Elementary Principles of Firearms; Elementary Principles of Ammunition; Powders and Explosives; Firearm, Cartridge and Bullet Markings; Instruments and Equipment; Preliminary Firearms Investigation; Advanced Firearms Identification; Digest of Court Decisions on Firearms Identification; Bibliography; Reference and Conversion Tabulations.

Following the title text there are 533 pp. of *Textbook of Pistols and Revolvers* reprinted in its entirety and bound in; for a description of which see # 18.

Frontis. of eight bullets which are described as having been fatal to their victims. Illustrated with B&W halftones, lines drawings, and sketches, some by the author.

SAMWORTH PROMOTIONAL: "All firearms experts and law enforcement officials are unanimous in their endorsement of this spendid textbook. There is no other volume like it. The author, an outstanding Ordnance Officer, was selected by the F.B.I. to instruct G-men in the identification, collection and presentation of proper evidence in cases involving firearms. This is the textbook used with a foreword written by J. Edgar Hoover. It thoroughly covers a highly technical subject in an interesting, readily understandable manner. Its complete text, tabulations and illustrations make it a worth-while (sic) and essential addition to the library of any collector, lawyer, detective or police official. It is a necessary bit of equipment to any law enforcement agency. Combined in one volume is the entire text of *'Pistols and Revolvers'* (sic) as the technical data it contains is highly necessary to any student. Some 900 pages, 300 illustrations."

SUBSEQUENT PRINTINGS AND ISSUES:

c1935 SATPCO Marines, Onslow County, NC. 2nd printing 1940.
c1935 SATPCO Plantersville, SC. 3rd printing 1942
c1935 SATPCO Plantersville, SC. 4th printing 1943
c1935 SATPCO Plantersville, SC. 5th printing 1946

The cases of all impressions are essentially identical. The type of paper used in the text varies greatly; the early printings having an uncalendered heavy book stock that is creamy in shade. The third through fifth impressions are printed on a grayish off-white, uncoated and calendered paper that has a look and feel not unlike newsprint. This change in stock was undoubtedly a result of wartime shortages and priorities.

SUBSEQUENT EDITIONS:

This work was considerably rewritten and published in 1957 under the same title by Stackpole Publishing with additional material by Frank J. Jury and Jac Weller, edited by Thomas G. Samworth. It does not contain *Textbook of Pistols and Revolvers* as a bound-in supplement, and is hence quite a bit thinner than the Samworth editions described above. The dust jacket carries the phrase "A Samworth Book", but it was not published by SATPCO, and is not considered a Samworth book by the compiler.

NOTES: Some first impressions of this title carry an errata sheet bound-in between the two books. The corrections pertain to errata that occur in the second book, *Textbook of Pistols and Revolvers.* A further discussion of this is presented separately in #18. The errata sheet will not be seen in examples of the stated second impression of 1940, as errata were corrected.

Dealers and collectors have noted confusion arising from the publication of the 1957 Stackpole edition which does not include the *Textbook of Pistols and Revolvers,* and the earlier impressions of the Samworth edition, all of which contain the second book. Samworth did serve as the editor for this 1957 Stackpole edition, which differs greatly from its predecessors with respect to editorial content, but it was not published as a Small Arms Technical Publishing Co. title. Therefore, as stated above, when this title is seen listed as having Hatcher, Jury and Weller as authors it is technically not a Samworth as defined herein.

The reader should note that this title is the only Samworth known to have the printing history of subsequent impressions listed on the title page verso.

20) Professional Gunsmithing
James A. Rogers Library collection

20) HOWE, Walter J.

PROFESSIONAL GUNSMITHING

A textbook on the Repair and Alteration of Firearms--with Detailed Notes and Suggestions Relative to the Equipment and Operation of a Commercial Gunshop

c1946 Thos. G. Samworth Plantersville, 1946 ad page date. Crown 8vo.

CASE: Navy blue heavily filled buckram. Gilt lettering on front and spine. Decorative gilt head- and tail-striping. "Samworth" in gilt above tailcap. Dust jacket illustrated with a Gayle Hoskins color painting of several men in a gunsmith's shop, one a grizzled hunter pointing out to the gunsmith a broken stock on his gun.

CONTENTS: 524 pp. including index, plus ads. Twenty chapters incl: Business Set Up and Customer Relationship; The Shop and Its Equipment; Special Tools and Fixtures; The Gun in for Repair; Function and Timing of Parts; Cleaning Actions and Removing Bore Obstructions; Adjustments and Considerations; Selecting and Mounting Sights; Making Small Parts; Woodwork, Stock Repairs and Alterations; Revolver Repairs; Automatic Pistol Repairs; Rifle Repairs; Shotgun Repairs; Heat Treatment of Small Parts and Springs; Grinding, Polishing and Blueing; Buying and Selling Used Guns; Specialty Work; Tables of Mechanical Reference; Souces of Supply.

Frontis. of dust jacket illustration by Gayle Hoskins. Illustrated with B&W plates, and numerous drawings by John B. Moll, Jr.

SAMWORTH PROMOTIONAL: "Gunsmithing methods and textbooks, like everything else, are matters of evolution and change for the better. This is particularly true of the many technical books on gunsmithing now in existence; each in part records or aug-

or augments the gradual but positive progress of the craft as time goes by.

"Here is a recent work written entirely from the professional outlook and devoted mainly to the repair and modification of existing stock weapons. *Professional Gunsmithing* approaches the subject from an entirely new angle, supplementing and enhancing all other previous Samworth Books on Firearms treating of this vital matter of gun repair and upkeep. It combines both business and technical phases of gunsmithing, in that matters such as time, ethics, price estimation, how to deal with shooters as customers, and the idea of doing the job exactly as someone else demands and not as the gunsmith himself might want to do it, are given paramount consideration throughout the text.

"How to best set up a gunshop and what is more important, how to keep it going on a profitable basis is included in this volume. Subjects such as business set-up and customer relationship are presented in clear, understandable manner. Several specialty lines which will provide a satisfactory source of revenue are included and the listing of sources of supplies for all gunsmithing materials is the most extensive yet published.

"Anyone entering the gunsmithing field, either as a means of livelihood or merely to work on their personal firearms, will find a mass of applicable material herein. The subject of commercial gunsmithing is taken up by Howe in a broad and comprehensive manner, approached from the angle of basic principles and logical reasoning. He tells how to diagnose gun troubles when the weapon is brought in for repair and explains the general reasons and causes for necessity of such repairs. By applying Howe's 'approach' to the problem in hand any gunsmith will be enabled to correctly and profitably remedy the fault or repair the defective part.

"Walter Howe's instruction is concise and practical because he knows whereof he writes. Specific jobs and types of repair work which have proved to be most frequently brought into the gunshop during his experience as a practicing gunsmith, are treated in detail. The amateur will appreciate this instruction as the advice of a master craftsman; the professional will be impressed with its clarity and practical application to the problems which daily confront him.

"*Professional Gunsmithing* is today's outstanding textbook for the gunsmithing profession, and is now being used as such in colleges and training schools which include gunsmithing in their curriculum. 520 pages, 120 special drawings, 21 plates."

SUBSEQUENT PRINTINGS AND ISSUES:

c1946 Thos. G. Samworth Georgetown, SC. 1952 ad page date. Crown 8vo. Case is similar to that of the first impression, but is a coarser weave royal blue filled buckram, without decorative head- and tail-striping. Dust jacket is illustrated like that of the first, with ads for later books, e.g. *Gunsmithing* by Dunlap.

SUBSEQUENT EDITIONS:

Later impressions using the original plates were made by Stackpole Publishing through the 1970s, but are dated only with the original copyright. Their dust jackets have the Hoskins illustration. Note: These printings often have a Samworth ad page that may include a few non-Samworth titles published by Stackpole.

NOTES: From a content standpoint, this book covers the many topics found in a gunsmithing book, some briefly and others in detail. The main thrust of the work is the gunshop as a business, and with this in mind the book well serves the intended purpose as a text for those formally studying to be a gunsmith.

Since it was considered a "working" book, *Professional Gunsmithing* is somewhat scarce in the first impression with a dust jacket in good condition. It is a reasonably common title, as Stackpole reprints are frequently seen on the used market. Samworth was usually quick to say that from a sales standpoint, his most successful books were the gunsmithing titles, with the Baker, Vickery, and Dunlap works leading the pack.

The compiler has seen two different dust jackets on the later 1952 impression, one with ads identical to those on the first printing jacket, and another with post-1946 titles listed. It is impossible to say with certainty that the 1952 impression was originally sold two different jackets; perhaps Samworth had several earlier ones left over and chose not to waste them.

It should be noted that Walter Howe was not in any way connected with James Virgil Howe of Griffin and Howe fame, author of *The Modern Gunsmith;* published by Funk and Wagnalls, various eds.

S.A.T.P.CO.

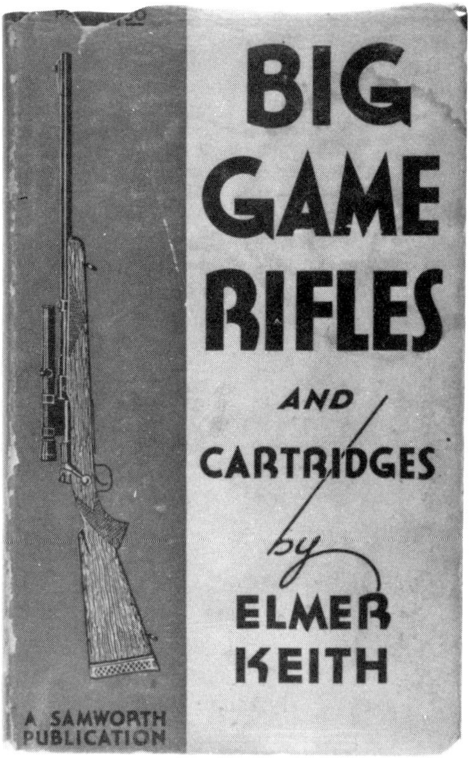

21) Big Game Rifles and Cartridges
B. R. Smith collection

21) KEITH, Elmer

BIG GAME RIFLES AND CARTRIDGES

c1936 Thomas G. Samworth Onslow County, NC. 1936 ad page date. 12mo.

CASE: Thin, medium-green pebble grain embossed cloth. Gilt lettering on front and spine. All edges green ink-washed. Dust jacket has two-color illustration of scoped rifle arranged vertically on the left, with title lettering on the right.

CONTENTS: 161 pp. plus ads. Foreword plus seven chapters incl: Brush and Timber Rifles; Long Range Stalking Rifles; All Around Rifles; Double Barreled Rifles; Iron Sights for Hunting; Hunting Scopes and Mounts; Hunting Rifle Stocks.

Frontis. of Keith and friend Ransome Henshaw with a mule deer "killed at long range". Illustrated with B&W halftone plates, and sketches by Philip Plaistridge.

SAMWORTH PROMOTIONAL: "A 170 page (sic), illustrated manual written by a man who has always lived amongst the big game of the Rockies and Western plains. Elmer Keith is outstanding as a rifleman, hunter and writer--and here in his first book he takes up our suitable big game rifles and cartridges one by one and tells just what they are capable of when put to the practical test of bringing down the game. A most original and accurate work, written by an author who is unique as a writer of shooting literature. All about brush and timber rifles--long range stalking rifles--doubles--all-around arms--and most practical chapters on iron and scope sights for the big game hunter."

SUBSEQUENT PRINTINGS AND ISSUES:

c1936 Thomas G. Samworth Plantersville, SC. 12mo. Case is similar to that of the first impression, but thicker, smooth fine-weave green cloth was used instead of pebble grain. 1943 ad page date. Dust jacket illustration is identical to that of the first.

c1936 Thomas G. Samworth Plantersville, SC. 12mo. Case is similar to that of the second impression, being a smooth fine-weave blue cloth. 1946 ad page date. Dust jacket illustration matches the earlier impressions.

SUBSEQUENT EDITIONS:

Facsimile reprinted in deluxe quarter-bound leather limited edition of 1500 copies by Wolfe Publishing Co. in the Wolfe Library Classics series.

Currently in facsimile reprint by Gun Room Press.

NOTES: This was the legendary Elmer Keith's first book, and is a title continuously sought after by collectors as such. Since the author's passing, demand has traditionally supported comparatively high pricing despite the availability of an inexpensive facsimile reprint. As with all the "Small Sams" with thin pebble grain boards, the "pebbling" has a tendency to flake off in the hinge area. Owners are cautioned against excessive opening, and the reprint is therefore recommended as a reading copy.

Collectors will note the smooth cloth bound second printing is somewhat thicker than the first; this is due to a higher caliper paper being used in the latter impression, a result of wartime priorities.

Keith stated in his second autobiography *Hell, I Was There!* that he wrote three "manuals" for Samworth; the first being *Sixgun Cartridges and Loads,* then *Big Game Rifles and Cartridges* and finally *Varmint and*

Small Game Rifles. Samworth chose to publish *Big Game Rifles and Cartridges* first, followed by the *Sixguns* title. Keith was advanced $250 for each of the three, and was to be paid an additional $250 each "if they sold", and he claimed the second payment on the first two books never came, although "(Samworth) later sold *Big Game Rifles and Cartridges* to the Stack-pole Company and it published a lot more of them." The compiler has not been able to verify the exist-ence of a Stackpole edition, but does not doubt author Keith's veracity.

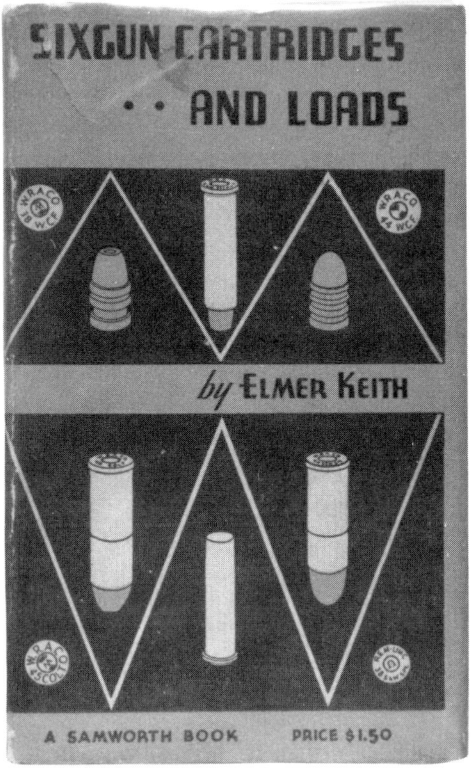

22) Sixgun Cartridges and Loads
B. R. Smith collection

22) KEITH, Elmer

SIXGUN CARTRIDGES AND LOADS

c1936 Thos. G. Samworth Onslow County, NC 1936 ad page date. 12mo.

CASE: Orange pebble grain embossed cloth. Gilt lettering on front and spine. All edges blue ink washed. Dust jacket illustrated with views of cartridges, headstamps and bullets arranged in a zig-zag pattern.

CONTENTS: 151 pp. plus ads. Foreword plus eleven chapters incl: Why We Reload for Our Sixgun; Handgun Cartridges and Their Possibilities; Selection of Bullets; Bullet Casting; Bullet Sizing and Lubrication; Revolver Powders; Cases for Reloading; Primers and Priming; Reloading Operations; Pressures, Primer Flattening and Case Expansion; Working Up Special Loads.

Frontis. of two photos of Keith holding a coyote and a string of jackrabbits, respectively. Illustrated with several B&W halftone plates, line drawings, and sketches.

SAMWORTH PROMOTIONAL: "This book pertains solely to the possibilities, selection and loading of revolver ammunition. It treats the special problems peculiar to the handloading of ammunition particularly suited to each of the various calibers and models of sixguns. The text goes into full details regarding minor and heretofore unmentioned facts regarding the essential components--bullet, powder, case and primer; to each of which is devoted full treatment in a separate chapter. A specialized work, covering all the many details and problems encountered in the assembling of accurate and powerful magnum loads; light gallery or practice charges; accurate target loads; long range loads; heavy defense cartridges; or the utmost in powerful, game killing cartridges loaded with specially designed hunting bullets. Each cartridge is taken up in turn, its full possibilities discussed and analyzed, then compared with other calibers in practical use and effect."

SUBSEQUENT PRINTINGS AND ISSUES:

c1936 Thomas G. Samworth Plantersville, SC. 1945 ad page date. 12mo. Case is similar to that of the first impression, but a smooth, fine weave orange cloth is used. Dust jacket illustration is identical to that of the first, but different ads appear on the flaps and back panel, including some post-1936 titles.

c1936 Thomas G. Samworth Plantersville, SC. 1945 ad page date. 12mo. Case is a medium blue buckram. Dust jacket is illustrated like that of the earlier impressions. It is believed that this is a later issue of the 1945 impression, since the ad page dates are the same, as are the books advertised on it. The paper stock appears identical, as does the ink lay of the type.

SUBSEQUENT EDITIONS:

Currently in facsimile reprint by Gun Room Press.

NOTES: Elmer Keith's second book, this title is a relative gold mine of information regarding handgun reloading and a "must read" for handgun enthusiasts. The major thrust of the content is dedicated to big bore handgun cartridges, and fans of the 44 Special and 45 Colt will want to include this work in their reference libraries.

It is vigorously sought after by Keith collectors; owners are encouraged not to frequently open their first impressions in order to minimize wear. The second printing and its later issue are slightly thicker than the first, although pagination is identical.

Samworth became acquainted with Keith while the former was editor of the *American Rifleman.* However, the Keith name did not appear on SATPCO titles until a decade after Samworth left the NRA; in the ensuing time Elmer became widely recognized as a firearms journalist. Samworth complemented Keith in a 1938 letter written to Townsend Whelen:

". . . (Keith's) writing is now considerably improved; his two books are doing nicely . . . the *Sixgun* book is outselling the rifle book almost two to one, but the *Rifle* book doesn't quite have the full measure of dope that is to be found in the other. The reloading chapters in *Sixgun Cartridges* are quite good; and I am certain that this simply the best practical revolver manual now out. Have one more manuscript of (Keith's) on varmint rifles, will wait a bit to see if the big game rifle manual sells or not as it is better than the varmint one."

If one reads between the lines, it can be inferred that Keith's third work, *Varmint and Small Game Rifles* was not produced because of lackluster sales on the part of #21, *Big Game Rifles and Cartridges*. For more on the *Varmint* title, see page 305.

The reader should note the use of the word "dope" in the letter above. This term appears in the promotional material for other titles and in this context does not refer to narcotics. During the first half of this century the popular definition was "information" but *Webster's* now lists that meaning fourth, behind 1) a preparation formulated to give a desired result, 2) narcotics, and 3) a stupid person.

23) .22 Caliber Rifle Shooting
M. L. Biscotti collection

23) LANDIS, C. S.

.22 CALIBER RIFLE SHOOTING

A Practical Volume Covering Target and Small Game Shooting with .22 Caliber Rifles

c1932 SATPCO Marines, Onslow County, NC. No ad page date. Tall 8vo.

CASE: Olive green silk embossed filled cloth with a decorative long-octagonal framed embossing of a woodchuck standing upright on a rock. Gilt lettering on front and spine. Decorative gilt head- and tail-striping. "STP Co." above tailcap. Oyster-white dust jacket illustrated by Ray Snow in single-color green ink showing a rear view of a hunter firing a rifle from the sitting position at a standing woodchuck at the far end of the field.

CONTENTS: 419 pp. Introduction and thirty-two chapters in two parts incl: Part I: The Small Bore Match Rifle; Ballistics of .22 Caliber Rifle Cartridg-es; Selecting Match Ammuntion; Range Selection and Construction; Sights and Their Adjustment; Target Telescopic Sights; Rifleman's Equipment; Wind, Light and Mirage; Shooting Positions; The Use of the Gun Sling; Coaching and Team Craft; Women and Juven-iles; A Talk on Accuracy; The National Rifle Asso-ciation; Famous Small Bore Rifle Shots. Part II: The Rifle on Game; Practical Hunting Trajectories; Out-door Trajectory Tests; Squirrel Shooting; Woodchuck Shooting; Crow and Hawk Shooting; Ducks and Geese; "Varmints"; Shooting Glasses; Practical Hunting Sights; Hunting with the Scope; the Boy's First Rifle; Cleaning the Rifle; Modern Hunting Ammuntion; Accuracy and Range; Choosing the Hunting Rifle; The Rifleman Afield. Two appendices include Small Bore Records and High Scores, and Small Bore Rifle-man's Directory.

Frontis. of competitiors on firing line at Sea Girt, NJ. Illustrated with B&W halftones. Ad follows text.

SAMWORTH PROMOTIONAL: "The first and only complete book devoted to the .22 rifle and its hundreds of thousands of users. It covers not only the rapidly growing Small Bore game but also is chock full of data for the many users of the .22 rifle in hunting. In fact, the author has specialized in the use of this caliber in the field and his book is crammed with interesting and practical facts relative to game shooting with the small bore rifle.

"A mass of highly original ballistic material is included, the chapters on trajectories being full of new technical data. All of the various .22 rifles are analyzed, from the heavy-barrel special match rifle for use at Sea Girt to the $5.00 single shot used by the boy in his first shooting experiences. All of the latest ballistic dope on the new Hi-Power .22 cartridges is available. And there is a most interesting and readable section describing some outstanding hunts with these cartridges. The squirrel and woodchuck hunters will find full details of their favorite sport, yet the book is fully as valuable and interesting to the fellow who merely carries a .22 when he goes for a walk in the woods.

"More than 400 pages together with a hundred illustrations taken especially to feature the text. Bound in standard silk cloth."

SUBSEQUENT PRINTINGS AND ISSUES: No known later impressions.

SUBSEQUENT EDITIONS: None.

NOTES: Outstanding text of the period that complements and somewhat resembles Crossman's earlier *Small Bore Rifle Shooting*, see #4. Landis details in this work much more information regarding the hunting use of .22 rifles as compared to the Crossman.

This is the scarcest of the three Landis books published by Samworth, and nowadays is almost never seen with a dust jacket. The market does not reflect this in that the latter two Landis titles are often quoted higher. The compiler believes one reason for this could be that many are under the mistaken impression that the content of *.22 Caliber Rifle Shooting* is outdated when compared to the later two works, or perhaps the public is under the erroneous belief that it is concerned only with the .22 rimfire cartridges. In reality, this book serves to describe the rifles and ammunition of the first thirty years of this century, and is a first-rate reference for firearms collectors interested in both rim and centerfire rifles of that period. Excellent and thorough coverage is given to the then-budding .22 wildcat cartridges like the Hornet, Lovell, and others, as well as the single-shot rifles built up for them. Whatever the case, it is an extremely collectible title, for Landis and Samworth enthusiasts alike.

This represents Judge Charles S. Landis' second book, the first being a small wrappers-bound 123 pp. pamphlet titled *Riflecraft*, c1923, The Sportsman's Digest Publishing Co, Cincinnati.

24) Twenty-Two Caliber Varmint Rifles
Dr. Paul Gilbert collection

24) LANDIS, C. S.

TWENTY-TWO CALIBER VARMINT RIFLES

c1947 SATPCO Plantersville, SC. Dec. 1946 ad page date. Tall 8vo.

CASE: Dark forest green silk weave filled cloth. Gilt lettering on front and spine. "SATP Co." in gilt at bottom of spine. Decorative embossed "frame" surrounds perimeter of front cover. Dust jacket illustration of a Gayle Hoskins color painting showing two hunters behind a stone wall; one taking aim with a rifle at two woodchucks in a distant field while his companion watches through field glasses.

CONTENTS: 531 pp. including index. Introduction and twenty-six chapters incl: History of .22 Varmint Rifles; The .22 Newton, .22/4000 and .220 Arrow; Commercial Varmint .22 Cartridges; The Lovell Cartridges; The Kilbourn Cartridges; The Lindahl Chuckers; Ackley Rifles and Cartridges; The Gebby Developments; The Blue Streak; The Donaldson Wasp; Canadian Developments; Niedner Cartridges; Some of "The Other Fellows"; Problems of Design; Selection of Actions; Tests of Modern Actions; Notes on Barrelmaking; Use of Chambering Reamers; Throating Chambers for Varmint Rifles; Bullets; Smokeless Powders; Preparation of Cases; Case Expansion and Fitting; Development of Special Loads; Tests and Opinions; Appendix.

No frontispiece. Illustrated with B&W halftones and line drawings in the text; there are no plates.

SAMWORTH PROMOTIONAL: "This is a book of 521 pages devoted exclusively to the increasingly popular .22 caliber "wildcat" rifles and their ammunition. Everything from the rebuilt single-shot action to the most elaborate custom built centerfire bolt action, good for 200 to 400 yard work on turkeys, woodchucks, crow, hawks, coyotes, and wolves, is treated in this extensive work. The best selection for any type shooting, every sort of neighborhood, every distance and every size pocketbook. All the good loads are listed, with their possibilities, their average accuracy at each relative distance, their killing power on different kinds of game and varmints, plus a wide assortment of loads for these various purposes. With this book, you don't have to guess at the right rifle, cartridge, loads or sights . . . or the type of hunting in which to use them.

"Here, for the first time, is presented full and correct technical information on the assembly, adjustment and finishing of most of the leading popular wildcat .22s of the day--boring, straightening and rifling of barrels; chambering problems, with special attention to the making of chambering reamers and their attendant and correct use; headspace adjustments; throating principles and processes; breaching and firing pin alterations and adjustments.

"An extensive and important section of the book is that devoted to commercial and military rifle actions for conversion into thoroughly safe and highly accurate varmint rifles. Most of the leading custom riflemakers of the United States and Canada have contributed their eminently qualified views on this vitally important subject, and have discussed it from every conceivable angle.

"The vital matter of high pressure .22 centerfire ammunition has been equally well handled; here are given dimensional specifications and essential loading data for all the popular .22 wildcats. The acquisition, conversion and preparation of special cartridge cases and bullets has been adequately covered: Here is the real dope on fire-forming; necking down; shortening necks; truing bases; reaming out case necks; primer pocket check-up . . . cartridge assembly such as never before presented to the handloaders.

Regardless of how much money you want to put into your varmint outfit, *Twenty-Two Caliber Varmint Rifles* tells all about that outfit. It tells about every well known wildcat cartridge which, to date, has been used sufficiently to prove itself. Every woodsman, every small game hunter, every farmer, rancher, stockman, professional varmint hunter or rifleman will find this to be the book he has long been looking for. Profuse with group reproductions, illustrations and cartridge drawings."

SUBSEQUENT PRINTINGS AND ISSUES:

c1947 SATPCO Georgetown, SC. 1952 ad page date. Tall 8vo. Case is dark forest green similar to that of the first impression, but a cloth of coarser weave is used without the decorative embossed "frame" around the perimeter of the front cover. Gilt lettering on front and spine. "Samworth" in gilt at bottom of spine. Dust jacket illustration is identical to that of the first. Jacket ads are different. Printed on heavier coated stock; this impression is approximately 1/4" thicker than the first printing.

SUBSEQUENT EDITIONS:

Reprinted by Stackpole Publishing in the 1960s, copies so marked.

NOTES: Lately popular due to the renewed interest in small caliber wildcat varmint cartridges, this is the most common Landis title of the three published by Samworth, owing to the number of later impressions. Text is somewhat a 'who's who' of .22 wildcats and wildcatters; it is an extremely thorough treatment of the subject from a rifle-cartridge-loading-gunsmithing standpoint and is of interest to any present-day .22 wildcat enthusiast. There is not much coverage given to hunting, this topic being treated at length in the author's later work, *Hunting With the Twenty-Two*, #25. The dust jacket illustration is a typically colorful Gayle Hoskins painting depicting two woodchuck hunters at work.

Used and rare book catalog listings and pricing reflect an apparent high demand, presumably for the intrinsic value of the text. This is a "hardware" oriented title of particular interest to enthusiasts of .22 wildcats and single-shot conversions, that is not particularly common despite the Stackpole reprints.

SAMWORTH BOOKS

25) Hunting With the Twenty-Two
Fred Talkington collection

25) LANDIS, C. S.

HUNTING WITH THE TWENTY-TWO

c1950 Thomas G. Samworth Georgetown, SC. 1950 title page date. 1950 ad page date. Short crown 8vo.

CASE: Medium tan unfilled coarse buckram. Dark forest green lettering on front and spine. Decorative head- and tailstriping of the same color. "Samworth" above tail cap. Dust jacket is illustrated with a painting by Gayle Hoskins of an autumn scene depicting a hunter taking aim with a rifle at a squirrel on a tree branch above a wooden fenceline.

CONTENTS: 429 pp. plus ads. Twenty-seven chapters, ten by others, incl: The Squirrels of America; Hunting Squirrels; Science in Squirrel Shooting With Rifles; Shotgun vs. .22 Rifle on Squirrels; Hunting Fox Squirrels; Hunting Black Squirrels (by John L. Cull); Last Day of the Squirrel Season; Bag Limit versus Long Runs; The Woodchuck; The .22 Long Rifle on Woodchucks; The .22 Hornet on Woodchucks; The Best Woodchuck Districts; Crow and Hawk Shooting With .22 Center Fire Rifles (by A. R. Weeks); The .22 Calibers on Hawks and Crows (by Henry E. Davis); Cartridge Selection for Crows and Hawks; Grouse Shooting With a .22 (by Bertram Chichester); Hunting Wild Turkeys With a Rifle (by Henry E. Davis); Wolf and Coyote Shooting (by C. E. Howard and Bud Dalrymple); Hunting Prairie Chickens (by E. B. Hutchinson); Fun With the .22 Rim Fire Rifle (by Henry E. Davis); The .22 in the North Woods (by Lloyd Melville); The .22 in Southern Swamps (by Fleetwood Lanneau); .22 Ballistics and Statistics; The .22 Hornet Cartridge (by C. S. Landis, Jr.); Telescope Sights; Elevation Readings for the .22 Long Rifle; Small Bore Rifles and Small Game Shooting of the Future.

No frontispiece. Illustrated with several B&W half-tone plates, with a few pen-and-ink sketches at the end of each chapter.

SAMWORTH PROMOTIONAL: "For years, the many devotees of .22 caliber rifles have been demanding an authentic and comprehensive work on the hunting of small game and varmints with these popular arms. Landis' *Hunting With the Twenty-Two* amply fills this need.

"Charlie Landis had spent more than fifty years in the active use of various .22 rifles: Twenty-twos of all types using the many different cartridges which have been available to American rifleman since the turn of the century. In this book he gives you the benefit of his lifetime of experience in the small game hunting field. Then, to supplement his own extensive personal experience, he has turned to those of his many friends who, through occupation or location, have been more fortunate in their opportunities to hunt and shoot larger or more inaccessible game with rifles of this caliber.

"The result is a book of 27 chapters comprising 425 text pages which are filled with instructive data and information on the hunting of the small game and varmints of the North American continent with the various .22 caliber rim and center fire cartridges. It covers all sections of these United States and Canada--Eastern hunting in heavily settled communities--Western experience in the open ranges and mountains--Northern trapping and hunting fields extending from British Columbia to the Hudson Bay country--and Southern shooting in the dense swamps and pine forests.

"The habits and histories of many game animals and birds are included, with much information as to the best means of hunting them. Here are many chapters on hunting the various species of tree squirrels--gray, black and fox; grouse, prairie chickens and ptarmigan with rifles; wild turkeys and other game.

Then come the varmints--crow, coyote, wolf and fox hunting; some lynx or wildcat are also treated and several chapters are devoted to that old standby--the woodchuck.

"All suitable hunting cartridges and loads for rim and center fire .22 rifles are covered with respect to their use in the hunting field--and when used as the main or as a supplementing rifle. Proper ammunition is discussed, as well as those pertinent factors of sight setting and aiming points.

"Here is a contribution of definite value to the riflemen of America--hunting information of a type not previously existent. *Hunting With the Twenty-Two* fully warrants the attention and reading of every owner and shooter of the more powerful, flat trajectory .22 rifles as well as the rim fire devotees.

"This is truly a practical book on hunting--not one man's theories or hearsay 'b'guess and b'gosh' stories. Well illustrated with numerous plates and drawings."

SUBSEQUENT PRINTINGS AND ISSUES:

No known later impressions.

NOTES: This popular work is the second-scarcest Samworth-published Landis title, but it enjoys the highest demand from collectors and shooters due to the subject matter. It is considered to be a companion of #24, *Twenty-Two Caliber Varmint Rifles*.

Hunting With the Twenty-Two is in demand for good reason; the content is everything Samworth said it was in his promotional copy, with the principal thrust on the practical use of such arms in hunting. Several of the chapters are written by others who were nationally and regionally known in the varmint hunting circle. With today's renewed interest in varminting and wildcatting, it would not be surprising to see both of these titles soon back in reprint.

The Gayle Hoskins-illustrated dust jacket, done primarily in autumn colors, is perhaps the most beautiful artwork on any Samworth title, in the compiler's opinion.

26) Big Game Hunting and Marksmanship
B. R. Smith collection

26) LEE, Kenneth Fuller

BIG GAME HUNTING AND MARKSMANSHIP
A Manual on the Rifles, Marksmanship and Methods Best Adapted to the Hunting of the Big Game of the Eastern United States

c1941 Thomas G. Samworth Onslow County, NC. 1941 ad page date. 12mo.

CASE: Dark royal blue pebble grain embossed cloth. Gilt lettering on front and spine. All edges blue ink-washed. Three-color dust jacket illustration by Alden Turner of a hunter crossing a log as he brings up his rifle to sight on a whitetail as it bounds away.

CONTENTS: 217 pp. plus ads. Foreword and ten chapters incl: Selecting Your Big Game Rifle; Hunting Sights and Telescopes; Hunting Marksmanship; The Rifleman in the Woods; Notes on Stalking; Small Game and Varmints; Notes on Guides and Hunters; Whitetail Deer and Their Hunting; The Black Bear; Eastern Moose and Their Habits.

Frontis. of a lake surrounded by Maine woods. Illustrated with B&W plates. Text stock is creamy-white.

SAMWORTH PROMOTIONAL: "A manual which treats solely of the arms, stalking methods and marksmanship necessary to bag the big game of these Eastern United States--being mainly devoted to the habits and hunting of the ever-popular white tail deer, with supplementary chapters on the moose, black bear and small furred game encountered by the average deer hunter. The author of this manual was a Maine guide for most of his adult life, who has hunted deer, bear and moose with practically all of the modern rifles and ammunition. He has spent more than 25 years in actual hunting, guiding and shooting and he presents in entertaining form the experience gained during that time. 217 pages and illustrations."

SUBSEQUENT PRINTINGS AND ISSUES:

c1941 Thomas G. Samworth Plantersville, SC. 1941 ad page date. 1941 title page date. 12mo. Later issue. Case is a dark green fine weave cloth, on boards slightly thicker than those on the first. Gilt lettering on front and spine. Spine lettering reads from top to bottom. Dust jacket illustration identical to that of the first.

c1941 Thomas G. Samworth Plantersville, SC. 1944 ad page date. 12mo. Later printing. Case is a royal blue fine cloth. Gilt lettering on front and spine. Dust jacket illustration identical to that of the first, but the ads are different. The text paper used is grayish-white and similar to newsprint, typical of wartime Samworths.

SUBSEQUENT EDITIONS: None.

NOTES: This book on hunting Eastern big game is one of the "Small Sams" and was intended to be a complementary book to *Big Game Rifles and Cartridges* published in 1936 and written by author Elmer Keith specifically for hunting Western big game. The Lee text was only moderately successful, having just a second impression in 1944. It is the compiler's belief that its limited success and lack of current demand is due to author Lee being relatively unknown compared to Keith, Landis, and others, although Lee had on occasion contributed articles to the *American Rifleman* while Samworth was editor. It is nonetheless an enjoyable and easy reading work of excellent content on Eastern big game hunting.

Collectors are cautioned not to frequently open first impressions as the pebble grain covers tend to crack and flake at the hinges.

27) Handloading Ammunition
Don Mason collection

27) MATTERN, J. R.

HANDLOADING AMMUNITION A handbook covering all phases of the loading of metallic ammunition for revolvers, pistols and rifles.

c1926 SATPCO Marshallton, DE. 1926 title page date. No ad page. Short 8vo.

CASE: Medium-brown lightly-filled coarse weave buckram. Orange ink lettering on front and spine. No dust jacket.

CONTENTS: 380 pp. Introduction and twenty-seven chapters in two parts: Part I: Handloading Ammunition; Tools For Handloading; Troubles and Cautions; The Loading Process Step by Step; Selection of Bullets and Their Fit; Jacketed Bullets; Solid Bullets; Bullet Metals, Alloys and Castings; Lubricating and Sizing Bullets; Powders For Pistols, Revolvers and Rifles; Measuring and Loading Powders; Ignition, Primers and Priming; The Cartridge Case; Ammunition Components, Their Cost and Where to Buy; Primary Ballistics For Handloads; Inspections, Tests, Markings and Records. Part II: Reduced Loads; Seating Bullets; Obtaining Maximum Accuracy; Handloading For Various Purposes; Quantity Handloading; Foreign Cartridges; Loading Obsolete Cartridges; Black Powder Cartridges; Handloading Cartridges For Revolvers; Suggestions On Popular Cartridges; Miscellaneous Information.

Frontis. halftone of a handloader's fully-equipped workbench. Illustrated with numerous B&W photos.

SAMWORTH PROMOTIONAL: "The reloading of central-fire rifle and pistol ammuntion pays. Rifle cartridges cost $3.50 to $12 per hundred rounds, depending on size. They can be reloaded as good as the factory product at from $1 to $2.50 per hundred. Moreover the average high-power rifle cartridge is suitable only for big-game or long-range target shooting, but the fired case can easily be reloaded with highly accurate loads suitable for the smallest game, or with loads for target practice in a gallery or at any range. You can make your own special cartridges for all purposes, fitting them exactly to your rifle or pistol so they will be more suitable and more accurate than the factory product and at one- quarter the price.

"Mr. Mattern's book tells exactly how to do this for any rifle, revolver or pistol, and in all calibers. It describes all the best tools, tells how to use them and where they can be had. Gives for the first time absolutely complete information as to rifles, cases, bullets, powder and primers which are necessary for either ordinary or expert hand-loading. It tells you how to get the utmost from your weapon in accuracy, power, range, efficiency and long barrel life and this with the utmost economy. Has complete information for every caliber and make of American weapons and many foreign. Where from other sources you get a paragraph or two on a certain rifle, in *Handloading Ammunition* you get five or six pages. The text, together with remarkable and original illustrations, make the instructions intensely practical and so plain that no novice can misunderstand. 380 pages, 117 illustrations, bound in buckram."

SUBSEQUENT PRINTINGS AND ISSUES:

Approximately 3000 copies were printed, and released in two issues, both identical. Samworth chose to bind only 2000 copies at first, and finished the binding in 1928.

SUBSEQUENT EDITIONS:

Facsimile reprinted in deluxe quarter-leather-bound limited edition of 1500 copies by Wolfe Publishing in the Wolfe Library Classics series.

NOTES: This is the first Small Arms Technical Publishing Co. title; the book that got the ball rolling for Samworth. After his resignation from the editorial position with the National Rifle Association in 1926, prompted by a difference in opinion with the organization's president, Samworth took the manuscript for this title back with him to the family property near Wilmington, DE. He knew J. Randall Mattern quite well, the latter having been Advertising Manager for the *American Rifleman*.

Samworth had originally intended *Handloading Ammunition* to be published by the NRA as a follow-up to its first hardcover book effort, *Amateur Gunsmithing* by Townsend Whelen. The latter was only marginally successful and NRA President Sandy McNab told Samworth "no more books!"

Mattern's book went on sale in late 1926 and was a moderate success; enough so that Samworth was encouraged to solicit further works from noted contemporaries as Townsend Whelen for *Wilderness Hunting and Wildcraft*, E. C. Crossman for *Small Bore Rifle Shooting*, Julian Hatcher for *Pistols and Revolvers & Their Use*, and Clyde Baker for *Modern Gunsmithing*. These books formed the foundation of the fledgling publishing company, eventually offering a total of forty-seven titles.

In correspondence, Samworth had this to say about the Mattern book: "Good seller, first one out and I made the price too low, this with the highest overhead has resulted in no money coming to (me) although I will make expenses and a bit over on it. Paid Mattern $650 to date on it with more coming."

Handloading Ammunition was considered the state-of-the-art reference on the subject at the time of publication. It contains much on the use of hand reloading tools, blackpowder loads, case preparation, bullet casting and sizing and the like, and it formed the basis for later works by others, notably Phil Sharpe and Earl Naramore. It is very scarce; the book did not come with a dust jacket.

28) A Rifleman Went to War
B. R. Smith collection

28) McBRIDE, Capt. Herbert W.

A RIFLEMAN WENT TO WAR Being a narrative of the author's experiences and observations while with the Canadian Corps in France and Belgium, September 1915 - April 1917. With particular emphasis upon the use of the military riflc in sniping, its placc in modern armament, and the work of the individual soldier.

c1935 SATPCO Onslow County, NC. No ad page date. Tall 8vo.

CASE: Fine weave lightly-filled cloth: copies seen in both canary-yellow and pumpkin-orange. Black art-deco lettering on cover; black lettering on spine, "STP CO." in black at bottom of spine. Dust jacket illustration by Alden Turner of scene of two infantrymen in WW I battledress; one of which is firing a rifle from a trench wall while the other spots the shots through a long telescope.

CONTENTS: 412 pp. plus ad page. Dedication and introduction with nineteen chapters incl: How Come?; Canada; England; Flanders; The Trenches; Record Scores; Scouting and Patrolling; Trench Raiding; Sighting Shots; The Pistol in War; The Battle of St. Eloi; Duds, Misfires and Stuck Bolts; The Somme; My Final Score; The British Army; Notes on Sniping; The Rifleman in Battle; The Machine Guns; The Soldier in Battle.

Frontis. plate portrait of author in uniform. Not illustrated.

SAMWORTH PROMOTIONAL: "This pioneer work on sniping and battlefield marks-manship (sic) was written by one of the outstanding American riflemen of all times, the late Captain Herbert W. McBride, whose name appeared high up on the lists of prizewinners at Sea Girt and Camp Perry for many years before the First World War. At the outbreak of that war, he enlisted in the Canadian Army and spent some 18 months in Belgium and France with the 21st Batallion, Canadian Expeditionary Force. The greater portion of this period found Mac (as he was known to thousands of us) actually in the front line trenches, with German infantrymen only a matter of yards away.

"His book, first published in 1936 (c1935), is still an applicable and authoritative work on the use of the modern military rifle and telescope on the battlefield. In World War II it was of incalcuable benefit to American and British forces in the training of their men in rifle marksmanship and the principles which McBride taught were very favorably commented upon by high ranking officers who observed and fought both in Europe and the Pacific.

"McBride served with Canadians as a machine gunner, his work as a rifleman being confined largely to sniping, though he used the rifle in repelling enemy attacks on more than one occasion and writes all about it. He carried a pistol, used it, and tells you what he thinks of it. He made something of a reputation on patrols and raids--having been decorated for work of this sort. There are chapters on each of these several subjects, dealing not only with his experiences but with his observations and conclusions. But the outstanding features are the chapters devoted to sniping and the rifle in battle. Captain McBride was one of the best qualified snipers in the Allied Army and he made a reputation in the Canadian Corps on his shooting ability. In this narrative, he goes into full details regarding the many tricks of the sniper--sniping equipment and its care and use--observation with the big telescope--concealment--range finding--counter sniping--and battle firing. No sob stuff and no dramatics--just a complete and truthful statement of the facts, telling you how to 'get your man' and get him 'fustest'.

"The instructions given, and the tricks Mac tells of, can be applied on any battlefield of today and in any static position where modern rifles and ammunition are used. His teachings are in no way obsolete; they still stand. Material and extracts from this book have been widely copied and quoted in many more recent domestic and foreign works and it is acclaimed as an outstanding contribution to our shooting literature. 412 text pages and frontispiece."

SUBSEQUENT PRINTINGS AND ISSUES:

c1935 SATPCO Plantersville, SC. No ad page date. Tall 8vo. Later impression. Case is identical to one of the issues of the first impression; being pumpkin-orange in color. Lettering on the cover and spine is gilt instead of black ink. The most obvious difference between the first and second impression is the change of publication location to Plantersville. Estimated to have been printed between 1941 and 1944, based on previous owner inscriptions found in copies examined, and books advertised on the dust jacket.

Dust jacket illustration is also the same as that of the Onslow impression; ads on the back and flaps are slightly different.

c1935 SATPCO Plantersville, SC. No ad page date. Tall 8vo. Later impression. Case is a medium-brown heavily filled smooth cloth. Black lettering on front and spine. Dust jacket illustration the same as that of earlier impressions. Believed to be a wartime impression due to the rough, grayish-white text paper in keeping with other books printed at the time, and from the titles listed on the ad page.

c1935 SATPCO Georgetown, SC. 1953 ad page date. Tall 8vo. Case is a heavily filled smooth brown cloth. Black lettering on front and spine. Dust jacket illustration is the same as that used on earlier issues but ads are for later titles. Dust jacket is printed on a semi-glossy coated paper, unlike the uncoated stock used for earlier jackets.

SUBSEQUENT EDITIONS:

Currently in facsimile reprint by Lancer Militaria.

NOTES: Outstanding personal account of WW I experience by an accomplished American rifle shot who volunteered in the Canadian Army. The title has been steadily sought after since its introduction in 1936. Of late, demand for the relatively scarce original Samworth impressions has led to the availability of a facsimile reprint introduced in 1987.

This was the first of four "wartime experience" personal accounts published by Samworth. Its success led to later post-WW II publication of Dunlap's *Ordnance Went Up Front*, #8; George's *Shots Fired in Anger*, #12; and Shore's *With British Snipers to the Reich*, #34. The four make a highly desireable collection in their own right, of particular appeal to enthusiasts of the military history genre.

Samworth had this comment regarding the McBride book in a 1933 correspondence: ". . . I also paid McBride $900 for a book on his experiences that will be a sure seller to riflemen when I get it out."

Captain McBride's work makes for very interesting reading, one time serious and sober in tone, the next fairly light-hearted. He has described all aspects of his personal experiences in the First World War, and gives freely of his opinions.

In one anecdote, the author relates an instance where the German 9mm Luger hit harder than a Colt .45 Automatic. The author had a small raiding party going through enemy trenches for German machine gun belts as they had earlier procured two such weapons but no ammunition. As the party made their way, they came across a German bent over a dead British soldier. A souvenir collector with McBride had time only to swing a recent acquisition, a holstered Luger, thoroughly coshing the German and laying him out cold. McBride notes this being the only time he was aware the Luger hit hard enough to get the job done.

Collectors should note most specimens of the Onslow and Plantersville impressions exhibit moderate to severe foxing on the frontispiece and the pages surrounding it. It is suspected that high acid content of the coated frontispiece sheet is responsible for the discoloration.

The author was a highly-respected rifle shot of considerable ability, and was a regular contributor to the *American Rifleman.* It was through this magazine that he made acquaintance with its editor, T. G. Samworth.

First impressions are seen with cases both canary yellow and pumkin orange in color. It is believed the yellow cases are on the first issue of the first impression, as orange cases are seen on both Onslow and Plantersville copies, suggesting the use of leftover binding materials on the later books.

29) Handloader's Manual
B. R. Smith collection

29) NARAMORE, Earl

HANDLOADER'S MANUAL
A treatise on modern cartridge components and their assembly by the individual shooter into accurate ammunition to best suit his various purposes.

c1937 Thomas G. Samworth Onslow County, NC. 1937 ad page date. 12mo.

CASE: Thin chocolate brown pebble grain embossed cloth. Gilt lettering on front and spine. All edges brown ink-washed. Dust jacket illustrated with drawings of cartridge cases, bullets and headstamps which are separated by a diagonal divider containing the title lettering.

CONTENTS: 369 pp. plus ads. Foreword and nineteen chapters in two parts: Part I - The Cartridge Case; Primers; Powder; Bullets. Part II - Bullet Alloys; The Casting of Bullets; Bullet Lubricants; Bullet Sizing and Lubrication; Mechanical Powder Measures; Powder Scales and Balances; Handloading vs. Ballistics; Handloading Operations; Revolver Ammunition; Loading for Automatic Arms; Loading for Extreme Accuracy.

Frontis. of graphical representation of the typical relationship between velocity and pressure for a given charge weight. Illustrated with B&W plates, and sketches by Lt. Col. Julian S. Hatcher.

SAMWORTH PROMOTIONAL: "Here is a different kind of textbook on the subject of handloading ammunition, one in no way resembling the conventional type of loading-tool catalog-handbook. It deals with the underlying principles of assembling ammunition in a manner that is applicable to any loading tool--past, present or future.

"For the first time, the individual shooter is instructed in cartridge nomenclature and the practice of ammunition assembly with the same thoroughness of detail and from the same angles that the technicians in ammunition factories are taught. The "what", the "why" and the "how" are given in full detail. Here is complete and explicit discussion and instruction, clearly explaining how to assemble dependable, accurate and safe ammunition. The subject is combined with that of interior ballistics and the performance of the cartridge is treated from the blow of the firing pin until the bullet reaches the target. As a result of this unusual treatment, the book is equally as valuable to the average rifleshot as to the most enthusiastic handloader. 375 (sic) pages. Specially illustrated."

SUBSEQUENT PRINTINGS AND ISSUES:

c1937 Thomas G. Samworth Plantersville, SC. 1943 ad page date. 12mo. Later impression. Case is a reddish-brown woven silk embossed cloth. Gilt lettering on front and spine. Dust jacket illustration same as that of the first impression.

NOTES: This is another of the "Small Sams" brought out as an inexpensive text that was affordable to most. Capably written over 50 years ago by an acknowledged expert in the field of ammunition, this title still has much to offer present-day handloaders.

While in the Army Naramore worked closely with such organizations as the Union Metallic Cartridge Co, Frankford Arsenal, Springfield Armory, Aberdeen Proving Ground, Lyman Gun Sight and others. Either *Handloader's Manual* or its successor, *Principles and Practice of Loading Ammunition*, #30, is an excellent reference for the advanced handloader.

The illustrations used in this title were reprinted from Mattern's *Handloading Ammunition*. The usual caution against unnecessary working of the hinges on the pebble-grain-bound first impression applies. First impression dust jackets are scarce.

Handloader's Manual

SAMWORTH BOOKS

30) Principles and Practice of Loading Ammunition
M. L. Biscotti collection

30) NARAMORE, Earl

PRINCIPLES AND PRACTICE OF LOADING AMMUNITION

c1954 Thomas G. Samworth Georgetown, SC. Dec. 1953 ad page date. Tall 8vo.

CASE: Red filled coarse buckram. Gilt lettering on front and spine. "Samworth" in gilt above tailcap. Dust jacket illustrated with Gayle Hoskins' self-portrait color painting "The Bulletmakers"; depicting a late-18th or early 19th century pioneer and his wife seated in front of their hearth casting round bullets. "A Samworth Book on Firearms" appears across the bottom of the dust jacket front panel.

CONTENTS: 952 pp. incl index; ads follow. Fifty-two chapters in two parts; Part I - Principles; The Cartridge Case; Parts of Cartridge Cases and Their Functions; Types of Cartridge Cases and Defects; Some Factors Affecting Case Life; Cartridge Cases and Arms vs. Safety; Primers; Functions of Primer Components; Kinds of Primers; Interchangeability of Primers; Ignition; Powder; Combustion; Those "Other Conditions" That Affect Burning Rate; Brief Description of Smokeless Powders; Selection of Powders; Powder Measuring; Powder Scales and Balances; Handling and Storage of Powder; Ballistic Measurements; Bullets; Sizing Cast Bullets; Bullet Alloys; Bullet Moulds; Jacketed Bullets; Reloading vs. Ballistics; Exterior Ballistics; Revolver Ammunition; Loading for Automatic Arms; Loading for Extreme Accuracy; Wildcat Cartridges; Keeping the Record Straight; Reloading Tools; Loading Dies, Chambers and Punches.

Part II - Practice: The Inspection of Fired Cases; Decapping; Cleaning Cartridge Cases; Operations on Case Necks; Resizing Cases; Operations on Case Heads; Seating Primers; Alloying Bullets; Bullet

Casting; Lubricating and Sizing; Homemade Jacketed Bullets; Seating Bullets; Powder Charging; The Estimation of Pressures; The Case Expansion Method of Estimating Pressures; Working Up Rifle Charges With Safety; A Guide for Working Up Rifle Charges; Tables of Revolver and Pistol Powders and Their Use; Making Measurements, Gadgets and Repairs. Appendix.

Frontis. of the Hoskins dust jacket illustration. Text illustrated with numerous B&W photos and pen-and-ink sketches, and printed on glossy medium weight coated stock.

SAMWORTH PROMOTIONAL: "Earl Naramore was born in Bridgeport, Connecticut, where for many years he was intimately acquainted and in close contact with personnel and officials of the Union Metallic Cartridge Company and the corporations that succeded it. Through these friends he became conversant with the manufacture of ammunition and the testing techniques of a ballistic laboratory. In 1924 he was commissioned a First Lieutenant in the Ordnance Officers Reserve Corps because of 'outstanding knowledge of ammunition'. As a Reserve Officer, he served at the Frankford Arsenal, Springfield Armory and Aberdeen Proving Ground. He was with the Lyman Gun Sight Corporation for six years, during which time he rewrote the Ideal Handbook, the text of which has been changed little since. In 1941 he went to Erie Proving Ground and for five years, until retirement, served as Chief Proof Officer, Chief of Inspection Division, Operations and Executive Officer.

"With this lifetime of study and experience in the commercial production of ammunition Earl Naramore has written the most comprehensive, accurate and applicable book ever published for the handloader of modern ammunition. Here is a basic and lasting work on the allied subjects of firearms, ammunition, cartridge components and ammunition assembly that will stand for years. It also clarifies the involved and not-too-well understood subject of interior ballistics. In this great work, Naramore has combined two books in one volume. Part One is a 565 page treatise on ballistic and mechanical factors and principles involved when a cartridge is fired in a gun--with the various components in the cartridge being taken up, and their specific functions and action clearly explained, singly and with relation to each other--with discussions of proper and improper methods of assembly by which the ballistics may be affected, even dangerously.

"This section goes deeply into the closely allied subjects of chemistry, metallurgy, explosives and physics--explaining in full all facts and ballistic principles which should be understood and applied if the individual is to produce reliable, accurate and safe rifle and revolver ammunition.

"Part Two outlines the mechanics of ammunition assembly. In this 350 page section various operations involved in the loading of the cartridge are completely discussed. This is not a loading tool catalog; with this book the reader will be enabled to turn out better cartridges with any loading tool. There are no generalities or vague treatment of any phase of handloading. Everything is written with a view of clearly explaining how the reader can excel in the production of high grade ammunition. It is original work--not based upon the writings or opinions of others.

"Fallacy of the average handloader in expecting or attempting to use old and obsolete tables of charges, or to be guided by the loadings of other individuals, in the assembly of high or extreme velocity cartridges, is clearly shown. Yet, the reader is told how to determine his own proper and safe charges with any brand or lot of smokeless powder--for any bullet--with any primer--in any rifle.

"Original and interesting material is given upon the methods and means by which the individual shooter can, with reasonable safety, take an unlabeled, unknown or uncertain brand of smokeless powder and, first: determine its suitability for use in the cartridge he has in mind and second: work up safe and proper charges of that powder.

"A new and logical method of enabling the individual to determine dangerous loadings by their pressure action and effect upon the brass cartridge case is also presented. This method, which Colonel Naramore personally developed and gives here for the first time, should be familiar to every cartridge experimenter or loader of wildcat cartridges.

"In this one book you are taught--WHAT should be done--HOW it should be--WHY it should be done that way. This thorough text is supplemented by 240 illustrations, many of which are micro-photographs (sic) taken by the author to clearly illustrate these teachings and principles. 915 text pages in all."

SUBSEQUENT PRINTINGS AND ISSUES:

No later Samworth impressions.

SUBSEQUENT EDITIONS:

c1954 Thomas G. Samworth, second printing published in 1962 by Stackpole and so marked. Case is a red leather-grain embossed filled cloth. Dust jacket is color illustrated identically to that of the first impression, including "A Samworth Book on Firearms" across the bottom of the front panel. "Stackpole" appears at the bottom of the spine on the jacket as well as the binding.

c1954 Thomas G. Samworth. Stackpole edition, third printing, undated. Case similar to that of the second impression. Dust jacket carries the same illustration as those of the previous impressions, and is marked "Third Printing" at the top of the front flap near the price area.

c1954 Thomas G. Samworth. Stackpole edition, fourth printing, so marked on dust jacket flap.

NOTES: Largely an updated and greatly expanded rewrite of the author's earlier *Handloader's Manual*, #29, this title has the distinction of being the last book published bearing the Samworth imprint.

It is an epic; 914 pages of text with more than 40 additional pages of appendix and index comprise this volume. Together with Sharpe's *Complete Guide to Handloading* (Funk and Wagnalls, var. years and eds.) it shared the position of preeminence as a definitive reference on the subject.

Although some data is outdated and some equipment referred to is obsolete, *Principles and Practice of Loading Ammunition* is a text that is aggressively sought by serious and advanced handloaders. It is not particularly scarce, having been reprinted by Stackpole as recently as the early '70s; these make excellent "working" references. Original Samworth impressions are understandably much less prevalent. It is not uncommon to see copies of this title to be in "shaken" condition, considering the 900-plus pages are printed on heavy coated paper.

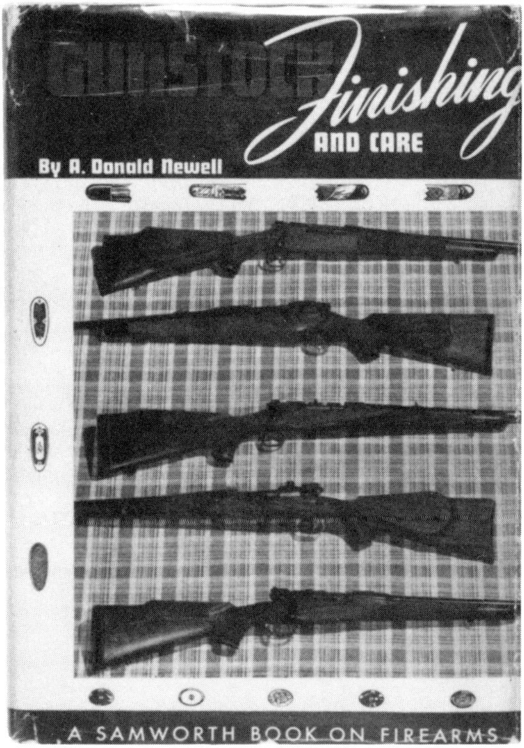

31) Gunstock Finishing and Care
B. R. Smith collection

31) NEWELL, A. Donald

GUNSTOCK FINISHING AND CARE

A Textbook, covering the various Means and Methods by which modern Protective and Decorative Coatings may be applied in the correct and suitable Finishing of Gun and Rifle Stocks. For Amateur and Professional Use.

c1949 Thomas G. Samworth Georgetown, SC. 1949 ad page date. 1949 title page date. Crown 8vo.

CASE: Grayish-tan lighly filled coarse buckram. Dark green lettering on front and spine. Dark green head- and tailstriping. "Samworth" in dark green above tailcap. Dust jacket illustrated with a photograph of five rifles arranged horizontally on the front panel and one vertical rifle on the spine panel.

CONTENTS: 444 pp. incl. index; ads. Introduction and sixteen chapters incl: Woods for Gunstocks; Preliminary Finishing Operations; Stains, Staining and Graining; Drying Oils and Their Application; Varnishes; Lacquers; Shellac; Plastic Finishes; Driers, Thinners and Solvents; Waxes, Polishes and Rubbing Compounds; Forearm Tips and Grip Caps; Refinishing Operations; Refinishing of Military Rifle Stocks; Antique and Early Gunstock Finishes; Auxiliary Equipment; Laboratory Tests and Notes. Six appendices follow: Natural Resin List; Evaporation and Drying time; Standard Liquid Measure and Equivalents; Glossary of Terms; Souces of Supply; Bibliography.

Frontis. of seven custom built rifles from the gunshop of W. R. Hutchings, Los Angeles. Illustrated with B&W plates and a few sketches. Tables of finish formulae are given.

SAMWORTH PROMOTIONAL: "This is an extensive and thoroughly complete contribution to the art of modern gunsmithing, a work equally valuable to either professional or amateur craftsmen. The author of this timely book is a technician in the laboratory of one of the largest American paint and varnish manufacturers; then, to further qualify in the preparation of so definitive a work he is a rifleman and amateur gunsmith of several years' standing.

"In this most thorough work Newell sticks entirely to his vocation in that he treats solely of gunstock finishing. In some 16 elaborate chapters, replete with heretofore unpublished trade data and technique, he tells everything known today by the people who make these finishing materials, so that they can be properly applied by the individual gunsmith seeking the most beautiful and thoroughly practical stock finishes for both modern and antique firearms. The greater portion of these procedures and formulae will prove virgin knowledge for the majority of our guncraftsmen.

"Herein is given authoratative and qualified information regarding all modern and early wood finishes which can logically be applied to gunstocks--as known to a professional technician and manufacturer. He gives the final word on bleachers, fillers, sealing compounds, water and moisture repellents, stains, driers, drying oils, varnishes, lacquers, shellacs and plastic finishes, with chapters also on waxes, polishing, cleaning and rubbing compounds. And, of course is included that old standby of the profession--the London Dull Oil Finish.

"Extensive chapters on primary and advanced treatments for all suitable gunstock woods outline selection, bleaching, sealing, graining, staining, waterproofing and all other necessary pretreatment data that will enhance or improve the natural beauties of the wood, plus an extensive assortment of all types of formulas that will enable the reader to select and compound his own finishing mediums, the like of which never previously have been available to shooters. More than 100 such formulae are given.

"The scale of treatments covered by this outstanding textbook ranges from the application of various solutions and solids by means of a rag or brush, or with pressure or spray gun systems, on through to the most modern technique of "baked finish" by means of a three-or four-tube bank of infrared lights.

"The text is written in non-technical form and all subjects discussed are presented simply and understandably in an interesting manner. Well illustrated throughout. 437 pages of text."

SUBSEQUENT PRINTINGS AND ISSUES:

c1949 Thomas G. Samworth Georgetown, SC. 1954 ad page date. 1954 title page date. Crown 8vo. Case is somewhat darker and more of a greenish shade of grayish-tan buckram, but the dust jacket is illustrated like that of the first impression. Jacket ads are for later works.

SUBSEQUENT EDITIONS:

Stackpole impressions have been seen, so marked; some of which are larger in format than the Samworth edition.

NOTES: That this title is concerned with such an ostensibly mundane subject accounts for its lack of notariety and relatively low demand. It is hard proof, however, that appearances can be deceiving.

Gunstock Finishing and Care contains a wealth of information on the subject of wood finishing in general, and of gunstocks in particular. This work is intrinsically valuable to gunsmith and cabinetmaker alike. The dozens of formulae presented yield unlimited possibilities regarding the staining and finishing of wood.

Both Samworth and Stackpole impressions are often seen with dust jacket, despite the title's "working" nature. Prices on the used book market are in general quite low and do not differentiate between editions.

32) Sixguns and Bullseyes
James A. Rogers Library collection

32) REICHENBACH, William

SIXGUNS AND BULLSEYES

c1936 T.G. Samworth Onslow County, NC. 1936 ad page date. 12mo.

CASE: Maroon pebble grain embossed cloth. Gilt lettering on front and spine. All edges red ink-washed. Dust jacket illustrated by Ray Snow; a diagonally-oriented title block separates two sketches: The top of the jacket shows a man aiming a revolver as viewed from behind, while the jacket below the title block has a sketch of a Smith & Wesson target revolver. "A Samworth Publication" is printed across the very bottom of the dust jacket front panel.

CONTENTS: 145 pp. plus ads. Thirty-four short chapters in two parts; Part I: Venerables of the Olymp; First Off; Taking Stock; Enthusiasm; What Do We Face; Guns and Headaches; Hold; Stance; Relaxation; Moving Into Position; Sighting; Squeeze; Breathing; Discourse on Habit; Resolved Therefore.

Part II: Ascent to the Olymp; Unity of Hold and Squeeze; Synchronized Sighting; Flinch-Wobble Company; Inhibitions and Ghost; Time and Rapid Fire; Adjusted Stance; Competition; Reputation; You and Your Gun; The Possible; Fun Galore; Some Kicks; Sights and Sighs; Trimming Your Gun; Grips; Ammunition Wrinkles; Gun Legislation; Finis.

Frontis. of the author with revolver being coached in a staged photograph. Illustrated with woodcut pen-and-ink sketches by Philip Plaistridge.

SAMWORTH PROMOTIONAL: "'The best handgun manual ever written.' Such was the verdict of the shooting clan when the first edition of this book (known as *The Elusive Ten*) came out in 1935. Its excellence and general application were such that the edition was soon exhausted. Here is the greatly enlarged and entirely rewritten second edition. Its author presents an entirely new and novel system in teaching the use of the revolver--a system which will enable any student to quickly become a first-class shot--or it will help the average shot to boost his scores above 83 and keep them there. New ideas--presented in a new style of writing--by a new author--and a good one. 151 pages (sic). Illustrated."

SUBSEQUENT PRINTINGS AND ISSUES:

c1936 T.G. Samworth Plantersville, SC. 1943 ad page date. 12mo. Case is dark red filled fine cloth, with gilt lettering on front and spine. Dust jacket illustration is identical to that of the first impression.

NOTES: This is another "Small Sam" which in the first impression has the fragile pebble grain case. The usual precaution against unnecessary working of the hinges applies. Second printings have a conventional thicker cloth binding and make excellent "reading" copies.

The Samworth Promotional material refers to the first edition of this work which appeared in 1935 under the title *The Elusive Ten*, published by Hathaway Oaks, which contained essentially the material found in Part I of *Sixguns and Bullseyes.* Samworth was so taken by the earlier work and its style of presentation that he commissioned Reichenbach to rewrite and enlarge it. A companion volume concerned with automatic pistols followed in 1937, *Automatic Pistol Marksmanship*, see #33.

The methods expounded by the author are of merit and interest to the handgunner. In the compiler's opinion, the only negative attribute to this book and its sequel is the unusual rambling style of prose employed by Reichenbach. However, it must have worked for revolver shooters of the post-Depression Era. Nothing succeeds like success.

Later printings are relatively common, and often seen in the dust jacket. First impressions are not all that rare, if one is persistent. This volume is usually priced somewhat lower than #33.

Sixguns and Bullseyes

SAMWORTH BOOKS

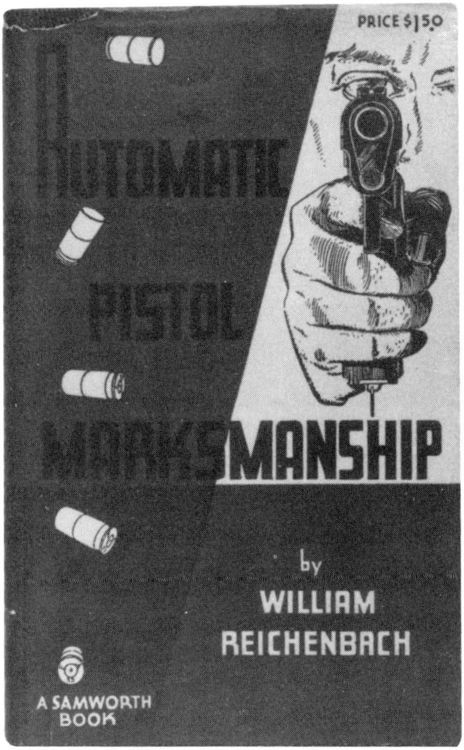

33) Automatic Pistol Marksmanship

James A. Rogers Library collection

33) REICHENBACH, William

AUTOMATIC PISTOL MARKSMANSHIP

c1937 T. G. Samworth Onslow County, NC. 1937 ad page date. 12mo.

CASE: Red pebble grain embossed cloth. Gilt lettering on front and spine. All edges red ink-washed. Dust jacket illustrated with sketch of a hand holding a 1911 Colt automatic pistol directly at the viewer, while five cartridge cases tumble down the left side of the front panel. Shooter's face visible behind gun.

CONTENTS: 140 pp. plus ads. Introduction and fourteen chapters incl: Automatics versus Revolvers; Material; Technique; Trigger Pains; Fluttering; Licking the .45; The Ideal Automatic; Ramblings; Holsters; The "Draw"; Preparing for the Fray; Ballistical Blotto; Homo Sapiens and Other Game; Allegretto. An epilogue, "Finale" follows.

Frontis. of a sketch showing a P.08 Luger superimposed above a battlefield surrounded by exploding shells. Illustrated with sketches by Richard Kroth and Edwin Bender.

SAMWORTH PROMOTIONAL: "Companion manual and continuation of the author's *Sixguns and Bullseyes*. Confined mainly to the use of automatic pistols and practical pistol shooting. The best of American and European self-loading pistols are described in detail and complete instructions are given for proper use of this modern type of handgun. Each individual model is described and analyzed separately. Then follows a most logical and sensible source of instruction teaching the individual the use of a handgun under conditions not to be found on the average target range. Quick draw--rapidfire--defensive--and similar types of practical shooting are discussed and taught in a manner entirely different from the methods heretofore employed. Illustrated."

SUBSEQUENT PRINTINGS AND ISSUES:

c1937 T. G. Samworth Plantersville, SC. 1943 ad page date. 12mo. Case is a thin, red fine cloth with gilt lettering on front and spine. Dust jacket illustration is identical to that of the first impression, but the ads are different.

SUBSEQUENT EDITIONS: None.

NOTES: Another of the "small Sams", this is the sequel to #32, *Sixguns and Bullseyes*, and deals with automatics in the same manner of treatment. The author describes numerous techniques for successfully mastering the use of auto pistols, and his methods indeed work.

First impressions employed the fragile thin pebble grain embossed case; a second copy for reading purposes is recommended. The second printing has a thin fine weave embossed cover which is not nearly as stiff as that used on the second printing of *Sixguns and Bullseyes*. Impressions are quickly identified by the location, Onslow for thc first and Plantersville for all later, as well as the different ad page dates which appear in both.

When one takes into ,account the date of publication, it is not hard to figure that several newer books on the same subject borrowed heavily from the methods and techniques Reichenbach presented herein. This title is a worthwile and inexpensive addition to the handgun enthusiast's library.

Automatic Pistol Marksmanship

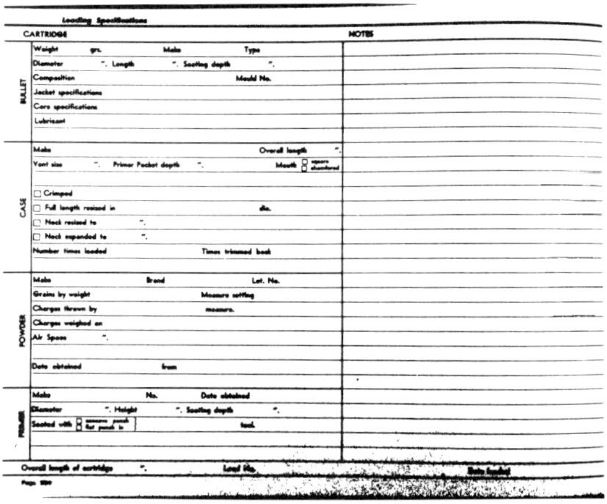
Loading Specifications

CARTRIDGE | NOTES

BULLET
Weight grs. Make Type
Diameter ". Length ". Seating depth ".
Composition Mould No.
Jacket specifications
Core specifications
Lubricant

CASE
Make Overall length ".
Vent size ". Primer Pocket depth ". Mouth ☐ square ☐ chamfered
☐ Crimped
☐ Full length resized in die.
☐ Neck resized to ".
☐ Neck expanded to ".
Number times loaded Times trimmed back

POWDER
Make Brand Lot. No.
Grains by weight Measure setting
Charges thrown by measure.
Charges weighed on
Air Space ".
Date obtained from

PRIMER
Make No. Date obtained
Diameter ". Height ". Seating depth ".
Seated with ☐ ☐ tool.

Overall length of cartridge ". Load No. Date loaded

34) Ballistic Records
James A. Rogers Library collection

34) SAMWORTH, T. G.

BALLISTIC RECORDS

c1953 Thomas G. Samworth Georgetown, SC. No ad page. Horizontal 4to.

CASE: Tan coarse-weave canvas. No dust jacket.

CONTENTS: 272 pp. of three types, all lined and ruled: 1) Loading Specifications, being the verso and providing the user with space to record ammunition data such as bullet weight, make, type, diameter, and composition; Cartridge case data such as make, primer pocket depth, length, neck diameter, number of times fired; powder data such as make, brand, lot, measure setting, charge weight, reference for reload information; and primer data such as make, number, lot, diameter, height, seating depth, etc.

2) Firing data, being the recto and providing space for recording the gun the cartridge was fired in, velocity, pressure, and comments such as recoil, report, primer appearance and case condition. Additional space is given for a record of the weather conditions, sight settings, bullet trajectory and point of impact, and general notes.

3) Firearm Specification Record, several pages in the rear part of the book for record of the shooter's rifles and their calibers, sights, weight, magazine capacity, stock dimensions and decoration detail, action type and make, barrel type, make, dimensions, rifling, etc.

Not illustrated. Introduction by Thomas G. Samworth.

SAMWORTH PROMOTIONAL: "The keeping of accurate and permanent records of all cartridge loadings and experiments is now a "must" for the handloader of modern ammunition. No longer can the individual safely prepare his cartridge upon the recommendations or opinions of other shooters or authorities or from the published Tables of Charges of years back. The high potential of present day primers and powders, which vary widely from lot to lot, permit no uncertainty as to the exacting assembly of proper components for any standard or high velocity load of today. Today's handloader must work up his own loadings--and remember them.

"*Ballistic Records* offers an extensive and permanent means of recording everything the keeper puts into his ammuntion, and also the definite results obtained from his varying loadings. Once a few score entries have been made, this book becomes a priceless bit of reference matter.

"Here is a large, stoutly bound book of 272 ruled pages, 8 1/2" X 10", prepared solely for the loader of modern ammunition, with space for 125 different cartridge loadings along with equal space to record their performance on the range or in the field, with forms equally applicable for rifle or pistol. There are eight different pages for the entering of all specifications, details and facts about the owner's rifles, no matter how minute or extensive these may be and it is an ideal form upon which to lay out the dimensions and lines of a prospective custom-built rifle.

"The target rifleshot has, for the past two generations, made a practice' of keeping his Range Score Book and closely plotting every shot fired at all ranges. Here, for the first time, is a similar and far more essential "scorebook" prepared for the use of the handloader and cartridge experimenter--compiled by a rifleman who has been using scorebooks and keeping loading records for the past 50 years.

"Substantially bound in heavy canvas. It should last for a lifetime--the longer it is used the more valuable it becomes."

SUBSEQUENT PRINTINGS AND ISSUES: None.

SUBSEQUENT EDITIONS: None.

NOTES: The only Samworth book with the publisher as author, this title has no text as such, but it contains the pages described above intended for use by the owner. In this sense, it is simply a notebook, to be consumed like any other. Hence, this title is of little interest to anyone except a serious Samworth collector. Used examples are sometimes seen in "junk boxes" at gun shows; as-new copies are virtually unheard of. This title is not at all a familiar one with used sporting book dealers. Prices vary widely.

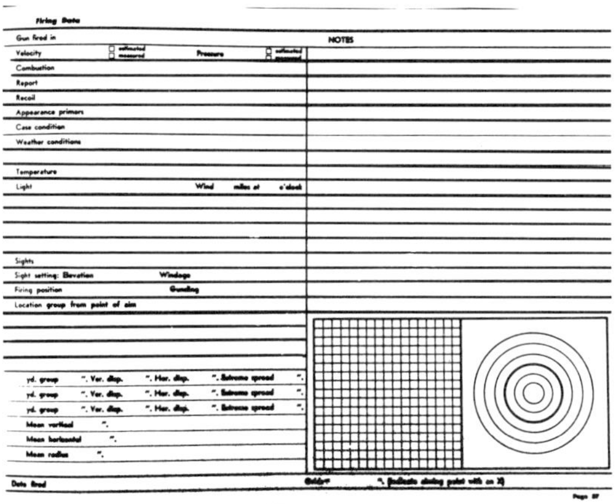
Firing Data
Gun fired in
NOTES
Velocity ☐ estimated ☐ measured Pressure ☐ estimated ☐ measured
Combustion
Report
Recoil
Appearance primers
Case condition
Weather conditions
Temperature
Light Wind miles at o'clock
Sights
Sight setting: Elevation Windage
Firing position Gunsling
Location group from point of aim
yd. group ". Ver. disp. ". Hor. disp. ". Extreme spread ".
yd. group ". Ver. disp. ". Hor. disp. ". Extreme spread ".
yd. group ". Ver. disp. ". Hor. disp. ". Extreme spread ".
Mean vertical ".
Mean horizontal ".
Mean radius ".
Date fired

35) With British Snipers to the Reich
Fred Talkington collection

35) SHORE, Capt. C.

WITH BRITISH SNIPERS TO THE REICH

c1948 Thomas G. Samworth Georgetown, SC. 1948 ad page date. 8vo.

CASE: Straw-colored unfilled coarse buckram. Dark brown lettering on front and spine. Decorative brown head- and tailstriping, with "Samworth" above tailcap. Dust jacket illustrated with an E. Stanley Smith painting showing a sniper with a 'scoped Lee-Enfield in a shell crater or foxhole. Title appearing on dust jacket is *With British Snipers* in lower case lettering.

CONTENTS: 351 pp. incl. index; ads follow. Fifteen chapters incl: Brief Odyssey; All Sniping; British Weapons and Equipment; Foreign Weapons; Pistols; Machine Guns; Ammuntion; German Weapon Training; Mainly in England; Hunting as Sniper Training; History of British Sniping; British Snipers in World War II; Miscellany; How a Sniper Was Made; One Man Sniping School.

Frontis. of the dust jacket illustration by E. Stanley Smith is titled "The Sniper". Illustrated with B&W plates.

SAMWORTH PROMOTIONAL: "World War II had been under way for some two or three years before British Army heads awoke to the obvious necessity for snipers, sniper training and sniper schools. But by the time the Rhine was crossed, the true value of trained and qualified riflemen was fully realized in all quarters and the British had sniper schools in operation in the various theaters of war--some only a few miles back from the battlefront. Although belatedly starting, once at it the British did a thorough job and fitted out their snipers with special rifles, binoculars, spotting scopes, camoflaged sniping suits and other essential equipment.

"With British Snipers is an accurate, thorough and interesting account of the organization, training and course of instruction at those various British Sniping Schools. The author went through the course at Zon in Holland and then served as an instructor until V-Day; previously he had landed in Normandy on D-Day and served throughout the advance across France and Belgium. He was also fortunate in training and serving with the famous Lovat Scouts, that banner outfit of the British Army when it comes to scouting and sniping.

"This is the most complete published work to date on sniping and sniper training. It fully covers the methods, rifles, telescopic sights and other sniping equipment issued in the British Army--as well as having considerable comment on similar material as used in the German Army. Range construction and range training are included. There is much interesting data on captured small arms and their ammunition. The matter treating of British small arms and ammunition is very complete and the author's comment on their Army and Home Guard training and marksmanship methods--and lack of same--is by no means the least interesting portion of this book.

"In addition to being a most qualified writer, C. Shore is that rarest of the genus homo to be found in the British Isles--a gun lover. He knew rifles and possessed a high degree of shooting skill before he entered the British Army--and this knowledge and skill is reflected throughout his book. Unfortunately, such qualifications are very, very rare in the British services--to their definite detriment. American riflemen will be astounded at what Shore charitably describes as the 'lack of firearms consciousness' experienced on all sides throughout his term of service, both at home and on the Continent.

"Yet our American forces went through the same war with even less interest in sniper training than the British at first possessed. We had very meagre suitable or special equipment and what little we finally got came too late to be of any use on the battlefield. Hence our American service commands can well afford to spend considerable time in reading and properly analyzing the contents of this book.

"Here is the latest word on sniper action and efficiency on the modern, mobile battlefield--in the face of automatic rifle, mortar-fire, and armored vehicle attack possibilities; of what can be done with the latest in sniper equipment and training; of the serious failings on the part of higher authority as to the value and proper utilization of that most efficient and economical man-killing machine--the modern, telescopic-sighted, sniping rifle with a trained rifleman in back of it. 350 pages, well illustrated."

SUBSEQUENT PRINTINGS AND ISSUES:

No known later Samworth impressions.

SUBSEQUENT EDITIONS:

1988 Lancer Militaria. Facsimile reprint of the Samworth.

NOTES: Without doubt this title is one of the rarest and most highly sought-after Samworths. Shore's book is complementary to McBride's *A Rifleman Went to War*, see #28; and the two join George's *Shots Fired in Anger*, #12; and Dunlap's *Ordnance Went Up Front*, #8 as the four "personal account" Samworths dealing with the subject of war.

All four titles were widely distributed through the post-WW II military, many ending up in post, unit or personal libraries. If an original Samworth impression is desired, *With British Snipers to the Reich* is more readily obtained by contacting used and rare book dealers that specialize in military subjects. More often than not, there will be a waiting list. The current reprint by Lancer is recommended as a "reading" copy; although the binding on the Samworth is strong, there is little sense in excessive handling.

For followers of E. Stanley Smith, his frontispiece and dust jacket painting is little-known; some may be surprised it exists.

There are few works on sniping available other than military training manuals issued by the major powers' armed forces training commands. Shore's book rightfully takes its place at the top of the general trade books on the subject. It is unique in that it concerns itself with philosophy and tactics rather than emphasize hardware like some works published in recent years.

To sum up, an original, as-new Samworth impression in dust jacket is a very rare find indeed.

S.A.T.P.CO.

36) The Gunsmith's Manual
Martin Beck collection

36) STELLE, J. P. and HARRISON, Wm. B.

THE GUNSMITH'S MANUAL

A Complete Handbook for the American Gunsmith, being of Practical Guide to all Branches of the Trade.

A facsimile reprint of the 1883 original.

Reprinted 1945 by Thomas G. Samworth Plantersville, SC. 1945 ad page date. 1945 title page date. 12mo.

CASE: Dark green filled woven silk embossed cloth. Gilt lettering on front and spine. "Samworth" in gilt above tailcap. Dust jacket illustrated with Gayle Hoskins painting entitled "Trade From the Monongahela" depicting frontiersmen outside a gunsmith's shop near Lancaster, PA; illustration wraps around the entire jacket outer panels.

CONTENTS: 376 pp. plus ads. Forty-two chapters incl: History of the Gun; How Guns are Made; Guns Now in Use; Pistols Now in Use; On General Gunsmithing; Taking Apart, Cleaning and Putting Guns Together; Tools Required for Work, Their Cost, etc; Tools, etc. and How to Make Them; The Work Bench; On Working in Iron; On Working in Steel; On Working in Silver, Copper and Brass; On Working in Wood; On Gun Stocks; On Gun Barrels; On Work on Gun Barrels; Tools For Breeching Guns; Tools for Chambering Breech Loading Barrels; On Gun Ribs; On Thimbles; On Rifling Guns; On Gun Locks; On Fitting Gun Hammers; On Nipples or Cones; On Springs; On Rods; On Bullet Molds; Screw Making Tools; Nomenclature; On Browning; Recipes for Browning; Miscellaneous; On Powder and Shot; Miscellaneous Recipes; On Judging the Quality of Guns; On Using the Rifle; On Using the Shot Gun; On Using the Pistol; Vocabulary of Mechanical Terms Used by Gun Makers; Vocabulary of Chemicals and Substances Used in Varnishes, etc; Calibres of Guns, Rifling, Twist of Rifling, Apart and Assembling Guns, Rifles and Pistols.

Frontis. of the Hoskins dust jacket illustration. Text illustrated with numerous woodcuts.

SAMWORTH PROMOTIONAL: "Back in 1882, when any town and village of any size whatever had its own hardworked gunsmith, J. P. Stelle and William B. Harrison wrote *The Gunsmith's Manual*--one of the very few early American works on firearms, and a book immediately recognized and accepted as the standard textbook of the gunsmithing profession. This publication held that position of eminence for several decades; in fact, it was the authority until the coming of smokeless powder.

"Stelle and Harrison produced a most authentic work thoroughly up to date for those times and exceptionally complete in that it treated upon all of the early breech loading firearms just then beginning to come into popularity and general usage. The book also was most thorough in its coverage and treatment of the muzzle loading firearms, which were then in their hayday.

"The long established standing of this early textbook, as well as its literary excellence, was such that we have reproduced it, page for page and word for word, as it was originally published.

"Despite the unique style and phrasing of its text, this is an enjoyable and at the same time a most instructive work. There is much amongst its material that is yet good workable shop practice; sound procedures and methods fully applicable by the gunsmith of today. Many of the formulae and bench tricks described herein can still be employed with profit. It fully covers all standard methods and processes followed in the days of muzzle loaders, both flint and cap-lock. In addition to clear and lengthy descriptions of popular fabrication methods, the book is profusely illustrated with most of the hand and bench gunsmithing tools necessary at that time.

"The average town gunsmith of the '80s and '90s made practically everything he needed except shotgun tubes, rifle barrels, and gun locks, and this book gives explicit directions, with illustrations, for the making of gun screws, nipples, keys, thimbles, hammers, ramrods and springs. Much data is given regarding the treatment, forging and processing of gun and barrel steel. The tools and processes necessary for the boring and rifling of those early gun barrels are described and illustrated. Quite a bit of text is devoted to cap-lock and early cartridge revolvers. Many are illustrated.

"This book is of considerable practical value to the present day gunsmith, as it is the only work in existence which treats authoritatively of the construction and repair of the muzzle-loading gun. Its extensive instruction relating to the fabrication of gun parts by means of hand tools and simple methods can also be of particular benefit to the worker with limited means at his disposal. 376 text pages with many illustrations."

SUBSEQUENT PRINTINGS AND ISSUES:

Case bindings on all copies examined are so similar as to be identical. The text and Samworth advertisements are likewise the same. Therefore, it is difficult to determine if more than one issue exists, due to lack of records regarding this title.

NOTES: On the surface an innocuous reprint of an earlier work, *The Gunsmith's Manual* characterizes the frustration and difficulty encountered when attempting to differentiate between first and subsequent impressions of a Samworth title. All copies examined appear to be identical, and there is nothing to suggest that different issues or impressions exist. Enthusiasts are encouraged not to worry about their copies of this particular title; better to congratulate the good fortune of owning a copy as it is somewhat scarce.

It should also be pointed out that the Gayle Hoskins dust jacket illustration "Trade From the Monongahela" was offered by Samworth as one of the two Samworth Prints on Firearms, which see in the respective chapter.

Historical firearms buffs and re-enactors will do well to acquire a copy of this title. There is much here for the blackpowder shooter, as the Samworth promotional material claims.

37) African Rifles and Cartridges
M. L. Biscotti collection

37) TAYLOR, John

AFRICAN RIFLES AND CARTRIDGES

The experiences and opinions of a professional ivory hunter with some thirty years of continuous living in the African Bush--who has used all of the various calibers and most of the suitable cartridges, and with them killed the many species of big game found on the continent of Africa.

c1948 Thomas G. Samworth Georgetown, SC. 1948 ad page date. Royal 8vo.

CASE: Gray-green lightly filled coarse buckram. Gilt lettering in maroon rectangular panels overprinted on front and spine; "Samworth" above tailcap in a similar maroon panel. Dust jacket carries a colorful E. Stanley Smith illustration of elephants among trees in the bush, headed toward waiting hunters.

CONTENTS: 426 pp. plus index. Foreword and fourteen chapters incl: Definitions and Details of Rifles; Double vs. Magazine; The Large Bores; The Large Medium Bores; The Medium Bores; The Small Bores; An All-Around Rifle; Sights, Sighting and Trajectories; Marksmanship in the Bush; Bullet Design and Construction; The Revolver or Pistol as Auxiliary; A Summing-up; Miscellaneous Odds and Ends; Afterthoughts.

Frontis. of rogue bull elephant and text sketches by E. Stanley Smith. Illustrated primarily with 25 B&W plates.

SAMWORTH PROMOTIONAL: "The 'big game' of today's world, the really dangerous game, is confined to the continent of Africa, plus a bit left here and there in Asia. Hence it is the dream of most of us American big game enthusiasts to some day have us the time of our life (sic) by hunting in Africa, which remains the true hunter's paradise, with animals of a size, temper and vitality unbelievable to the average sportsman who has never hunted there or studied of its varied and plentiful fauna.

"Furthermore, Africa is still a place where a good shot can earn his livelihood with the rifle. John Taylor, the author of this book we are about to describe, has done so for thirty years; he is now back there in the bush living off his rifles. For these past three decades he has been a professional ivory hunter, one who has spent from eleven to twelve months of each year in actual hunting. During these years he has killed more than a thousand elephant, a far greater number of buffalo, countless lions, hippos, rhinos, crocs, antelope of all sorts, and all the other queer critters which inhabit that continent. And he is doing it today, hunting mostly in Portugese East Africa, although his past travels and safaris have taken him over all of the African game fields as well as trips to India, Australia and North America.

"Some two years back, we approached John Taylor with a request to write us a book covering this lifetime of African hunting, but a different sort of book from anything previously written. We did not want a 'hunting book'; we wanted a work donc entirely from the standpoint of the rifles and ammunition used, one written from the viewpoint of a skilled, experienced hunter who knew the practical ballistics of rifle shooting as well as the mere ability to point a rifle and hunt correctly. Taylor has for years been a close reader of the sporting press as well as an observer and experimenter of all the modern improvements in rifles, ammunition and ballistics. In this book he gives us the peculiar combination of a professional hunter who has been able to back up his daily hunting with the latest in correct ordnance and ammunition development and knowledge, and to tell us how it all worked out.

"In *African Rifles and Cartridges* John Taylor takes up the modern and powerful British, German and American big game rifles and cartridges and, one by one, tells all about them. And he is fully qualified to do so, as he has used them all in his thirty years of professional hunting. Doubles, magazines and single-shots, he can and does tell of the virtues and faults of each. He tells of the practical value of steel jacketed bullets versus those covered with cupro-nickel; he points out the faults of copper-tubed bullets; he shows the actual effect of soft-point bullets with varying degrees of lead exposure; he explains the necessity of modern bullets that will really stand up to the requirements when fired into heavy animals; as well as numberless other tips and explanations about cartridges and rifles. Here is ballistic information of a degree unknown heretofore to us American hunters; information that we can profitably apply in the hunting of our native big game.

"Supplementing all this practical application of ballistics is a wealth of information on the actual hunting of all African game; the habits of many species; together with instruction on the practical handling and shooting of the different rifles under discussion. Gunbuilding specifications, stocking, weight, balance and fit are fully discussed, as well as the true worth of the many accessories and gadgets. Sighting arrangements are all here; the author is a strong advocate of modern telescopic sights and he uses them whenever possible, even when hunting the most dangerous game. One outstanding chapter is 'Marksmanship in the Bush', a positively unique exposition of the practical application of one's shooting ability when faced with the real test in the hunting field.

"This outstanding work is supported by a string of technical illustrations, especially prepared and as original as the text they clarify. The many cartridges, their components, and the various methods of ammunition assembly employed by gunmakers of different countries are all clearly shown in full-scale detail and by a new method of depiction, one you will clearly understand and appreciate. Our staff artist, Ned Smith, worked out a highly original scheme in this arrangement of cartridge illustration, which he presents for the first time to American shooters.

"Any rifleman or big game hunter will enjoy and profit from reading this book. 426 pages, with technical drawings and illustrations galore."

SUBSEQUENT PRINTINGS AND ISSUES:

c1948 Thomas G. Samworth Georgetown, SC. 1952 ad page date. Royal 8vo. Except for the ad page date, case and dust jacket are identical to that of the first impression.

SUBSEQUENT EDITIONS:

At least two Stackpole impressions exist. The earlier Stackpole impression has a case similar to that of the Samworth impressions, complete with lettering in an overprinted block of contrasting colored ink; the later Stackpole omits these blocks, with the gilt lettering applied directly on the cover.

Fascimile reprinted in a deluxe quarter-leather-bound limited edition of 1500 copies by Wolfe Publishing Co. in the Wolfe Library Classics series.

Also available in facsimile reprint by Gun Room Press.

NOTES: An ever-popular African title by a well-known and respected author. Due to the subject matter this is one of the most sought-after and highly-priced Samworth titles; it is moderately scarce in the Samworth edition issues. Some Stackpole reprint editions have been seen advertised at pricing levels to match the Samworths. These Stackpole impressions are physically nearly identical to their SATPCO predecessors, and some used and rare dealers may not be aware of the difference. The availability of the

Wolfe Publishing and Gun Room Press reprints has taken some demand pressure off this title from those seeking "reading" copies.

For anyone interested in the subject, this text is at oncc informative and enjoyable. Rigby rifle and cartridge enthusiasts will also find it of interest as Taylor comes across clear in his preference of these fine arms.

Riling in his bibliography *Guns and Shooting* quite correctly points out that *African Rifles and Cartridges* is "a revision of and extension of material" presented in a 215 pp. work entitled *Big Game and Big Game Rifles* published concurrently in England by Herbert Jenkins, Ltd.

It is difficult to determine with certainty whether the later Samworth impression is truly a printing or merely a later issue with new ad pages tipped in. Dust jacket ads are different on the later printing, but it is impossible to discern a difference from an examination of the text paper and the ink lay of the print. One should probably not be too concerned about points of an issue, owing to the title's scarcity.

Although quite scarce in the Samworth issues, this is perhaps one of the the most widely reprinted SATPCO titles, a testimony to its timeless appeal.

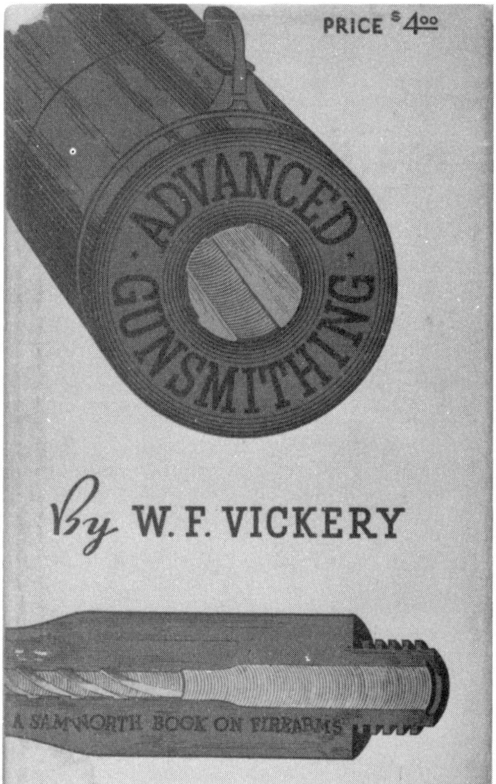

38) Advanced Gunsmithing
B. R. Smith collection

38) VICKERY, W. F.

ADVANCED GUNSMITHING A Manual of Instruction in the Manufacture, Alteration and Repair of Firearms in-so-far as the Necessary Metal Work with Hand and Machine Tools is Concerned. With Chapters on the Boring, Rifling and Chambering of Barrels. For Amateur and Professional Gunsmiths.

c1940 Thomas G. Samworth Onslow County, NC. No ad page date. 1940 title page date. 12mo.

CASE: Medium brown pebble grain embossed cloth. Gilt lettering on front and spine. All edges brown ink-washed. Dust jacket illustrated with a gun barrel muzzle around the bore of which appears the title. The chamber area of a sectioned rifle barrel occupies the lower quarter of the jacket front panel.

CONTENTS: 428 pp. plus ads. Twenty chapters incl: Shop Equipment; Barrel Changing and Its Adjustments; Chambering, Boring and Reaming Tools; Rifling Tools and the Rifling of Barrels; Reboring and Rechambering Old Rifles; Action Work and Alterations; Sights, Scopes and Small Parts; Shotgun Repairs; Problems of .22 and Other Rim-Fire Rifles; Revolver and Automatic Pistol Jobs; Cleaning, Clearing and Lapping Barrels; Working with Hand Tools; Notes on Firing Pins; Soldering and Brazing; Forging and Welding; Heat Treatment of Steel; Blueing of Firearms; Cartridge Case and Bullet Dies; Loading Tools, Accessories and Appliances; Postmortem. Several pages of sources of supplies and information follow.

Frontis. of trigger pull adjustment viewed through a magnifying glass. Illustrated with line drawings and sketches by Oliver B. Hamilton.

SAMWORTH PROMOTIONAL: "*Advanced Gunsmithing* is a specialized manual, restricted to the metal-working phases of gunsmithing. In these various operations it takes the reader many stages farther than do earlier works, yet at the same time it is a most essential book for the amateur to have at hand. This is a book we can conscientiously recommend as an early purchase by the beginner in firearm study and repair.

"W. F. Vickery is a professional gunsmith of long experience, and is also a gun-crank and experimenter of advanced degree--one who fully understands the complex ideas and requirements of the modern rifleman. Practicability is the keynote of his text, and his methods and instruction will prove workable and applicable under the conditions confronting the average owner of firearms. You can put his text to practical application on your guns--right there in your own home.

"Here are given principles of and complete instruction on barrel making, boring, reaming, rifling and chambering--done by means of the small modern power tools now so popular, or done entirely by hand on a hand-rifling bench using rifle-heads, barrel and chamber reamers which you yourself can make. It tells all about headspacing, barrel changing, action alterations, barrel lining, together with some highly applicable chapters on the reboring and chambering of rusted or shot-out barrels or of obsolete rifles. Extensive treatment is given the popular .22 rimfire and its peculiar problems. There is a full chapter on shotguns and their repairs or alterations. The causes and corrections of malfunctions are given. Complete coverage is made of all metal small parts: sights, barrel bands, swivels, firing pins, lock and action parts, telescopic sight fittings and repairs, coil and flat spring making, heat treatment and tempering of gun parts. Comprehensive treatment is given of all the small repairs most frequently necessary in firearm upkeep.

"No matter how many other gunsmithing books you now have, you need this one. It takes up where all others left off. 432 (sic) profusely illustrated pages filled with data on modern firearms and the most applicable and modern methods of adjustment and repair."

SUBSEQUENT PRINTINGS AND ISSUES:

c1940 Thomas G. Samworth Plantersville, SC. 1943 title page date. 12mo. Later impression. Case is a light medium brown pebble grain embossed cloth, all edges brown ink-washed. Dust jacket is illustrated like that of the first impression.

c1940 Thomas G. Samworth Plantersville, SC. 1943 title page date. 12mo. Later issue of the second impression above. Case is a medium brown vellum-finish smooth cloth, all edges brown ink-washed. Dust jacket is illustrated like that of the first impression, but ads are different.

c1940 Thomas G. Samworth Plantersville, SC. 1946 title page date. Later impression. Case and dust jacket as (1943) above. Case is a medium brown vellum-finish smooth cloth, all edges brown ink-washed. Dust jacket is illustrated like that of the first impression, but post-1944 titles are listed.

c1940 Thomas G. Samworth Georgetown, SC. 1951 title page date. Later impression. Case and dust jacket as above, except jacket ads are different, and the text is slightly thicker due to heavier basis weight stock.

c1940 Thomas G. Samworth Georgetown, SC. 1955 ad page date. Probable later issue of 1951 printing; text paper and print appearance virtually identical. Case and dust jacket as (1951) above.

SUBSEQUENT EDITIONS:

c1940 Thomas G. Samworth Harrisburg, PA. No ad page date. 8vo. Stackpole facsimile reprint. Larger format; has light tan coarse weave buckram case, with brown lettering on front and spine. Has undated Samworth-type ad page that lists non-Samworth titles.

Facsimile reprinted in a deluxe quarter-leather-bound limited editon of 1500 copies by Wolfe Publishing Co. in the Wolfe Library Classics series.

NOTES: An extensively printed Samworth title that enjoys continuing steady demand from gunsmiths, both amateur and pro alike. Since the first printing has the fragile thin pebble grain covers, enthusiasts are cautioned to refrain from excessive working of the hinges to prevent flaking of the cover material.

This work is well-known for its content; it delves deeply into advanced metal working techniques employed in the gunsmith trade, much more so than the Baker or even the Dunlap works. It enjoys a steady demand from customers of the used and rare book market.

Collectors should take note of the Stackpole reprint; it is fully three-quarters of an inch taller than the Samworths, and contains an ad page listing "Samworth Books on Firearms" that list non-Samworth titles as well. Stackpole used these lists to market remaining inventory after Samworth sold out to them.

Copies on the used market are rarely seen in dust jacket, particularly the first impression, since this was a "working" title moreso than a shelf-sitter.

S.A.T.P.CO.

39) Wilderness Hunting and Wildcraft
B. R. Smith collection

39) WHELEN, Townsend

WILDERNESS HUNTING AND WILDCRAFT

With notes on the habits and life histories of big game animials.

c1927 Small Arms Technical Publishing Co. Marshallton, DE. No ad page. 8vo. Dust jacket: See NOTES.

CASE: Dark blue silk embossed filled cloth. Gilt lettering on front and spine. Decorative gilt head- and tailstriping. "SATP CO." in gilt above tailcap. Decorative embossing of moose skull with horns on front cover.

CONTENTS: 338 pp. of text. Introduction and twenty-four chapters incl: The Habits of Big Game; The Three North American Deer; The White-Tailed Deer or Virginia Deer; The Mule Deer or Rocky Mountain Black-Tailed Deer, Jumping Deer; The Columbian Black-Tailed Deer; The Wapiti or Elk; The Moose; The Caribou; The Mountain Sheep; The Rocky Mountain Goat or White Goat; The Bears of North America; The Big Brown Bears; The Grizzly Bears; The Black Bears; Still-Hunting or Woods Hunting; Stalking, or Mountain or Plains Hunting; Finding One's Way; Clothing and Personal Kits; Camp Beds and Bedding; Shelter and Tents; Big-Game Rifles; Wilderness Marksmanship; Photography; Physical Preparation. Three appendices follow: Appendix A, Convenient Checklists; Appendix B, Additional Data and Information; Appendix C, Bibliography.

Frontis. photo of a bighorn sheep. Text illustrated with B&W photo plates and sketches by the author.

SAMWORTH PROMOTIONAL: "Are you aware that about sixty percent of the big game obtained by the average sportsman is killed by the guide? The average sportsman fails generally by his lack of knowledge of hunting and shooting. The sportsman who always follows a guide has no comprehension of what hunting--that is, still hunting or stalking combined with woodcraft--means. Avoid such humiliating experiences on your next hunt.

"This is the only book which really teaches the hunting and shooting of big game. It is entirely different from anything heretofore published. The author has had a lifetime of experience in wilderness hunting, largely depending on his own efforts. *Wilderness Hunting and Wildcraft* tells in a most practical manner just how to enter a wilderness, how to find one's way, discover where the game is, still hunt or stalk it, and how to bring it to bag or photograph it. It also makes a strong plea for conservation and good sportsmanship. The book is also valuable to guides, explorers and all who have to live by their rifles in a strange country.

"Contains chapters on the habits and hunting of each species of our big game which are full of interesting anecdotes and experiences of the author or our leading big game hunters. Chapters on the wildcraft which the hunter needs in woods, mountains or plains; on equipment for all countries; on rifles; wilderness marksmanship, wilderness photography, and physical preparation. The book and its illustrations are entirely original throughout, and full of information not to be found elsewhere. 340 pages, 75 photographs, and numerous sketches and diagrams. Handsomely bound in cloth."

SUBSEQUENT PRINTINGS AND ISSUES: None.

SUBSEQUENT EDITIONS:

Facsimile reprinted in a deluxe quarter-leather-bound limited edition of 1500 copies by Wolfe Publishing Co. in the Wolfe Library Classics series.

NOTES: Very scarce title by a very collectible author. This book should be in the library of any serious North American big game hunter, as Whelen was predictably ahead of his time when he wrote it in 1927. Some text material was originally published in 1923 in one of the small 100-page books in the "Outers Recreation" series entitled *Big Game Hunting.* After being approached by Samworth, Whelen rewrote and expanded the 1923 work, more than tripling its content. An immensely enjoyable book, *Wilderness Hunting and Wildcraft* brings to the reader the essence of the total outdoors experience. The author was an advocate of packing in one's necessary equipment and being totally self-sufficient while on the hunt. This work contains a wealth of pragmatic advice along those lines that is seconded by few other works.

As a Marshallton book and therefore an "early" Samworth, it is relatively scarce, being the second title to appear under the Small Arms Technical Publishing Co. imprint. The compiler has never seen a copy in dust jacket, it may not have been issued with one. An effort to locate one for photographing was fruitless.

Samworth was disappointed with the flat sales of this title, which were responsible for there being only one printing: "After two years the Whelen book is almost a flop; not enough 40-year-old Boy Scouts in the USA to make it worthwhile. Still have half the edition on hand with sales very slow, doubt if the entire edition sells out in the next ten years. I have my money back out of it now, and will even make some, but slowly."

Since slightly less than 3000 copies were printed, enthusiasts should take advantage of an opportunity to acquire it. Pricing runs from inexpensive to very high, depending on condition and geography; other Whelen works are better known.

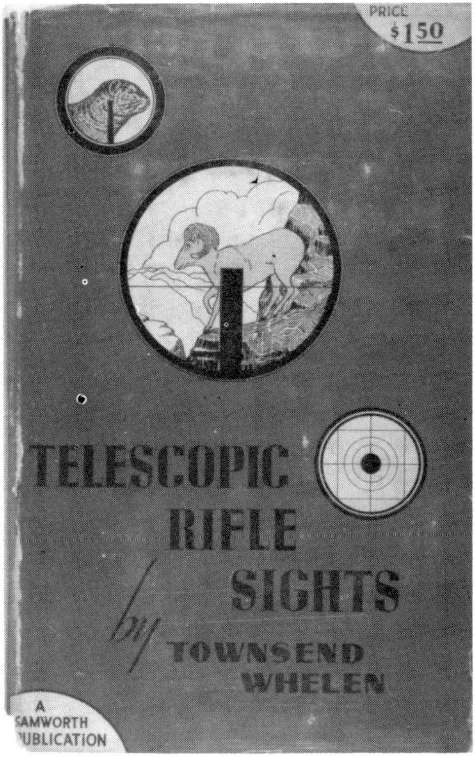

40) Telescopic Rifle Sights 1936 Edition
B. R. Smith collection

40) WHELEN, Townsend

TELESCOPIC RIFLE SIGHTS

In two editions: c1936 Onslow County, NC; c1944 Plantersville, SC. (Revised and expanded)

FIRST EDITION:

c1936 Thomas G. Samworth Onslow County, NC. 1936 ad page date. 12mo.

CASE: Black pebble grain embossed cloth. Gilt lettering on front and spine. All edges dark blue ink-washed. Dust jacket is line-drawing illustrated with views through a riflescope of a woodchuck, a bighorn sheep, and a target bullseye respectively. "A Samworth Publication" appears in a semicircle at the lower left front corner of the jacket.

CONTENTS: 131 pp. plus ad page. Six chapters incl: General Principles; Target Scopes; Small Game and Varmint Scopes; Big Game Hunting Scopes; Junior Scopes; Optics of the Scope. A one-page directory of scope manufacturers follows.

Frontis. of a Springfield rifle fitted with a Noske scope. Illustrated with B&W plates, with sketches and line drawings by the author.

SAMWORTH PROMOTIONAL (1936 ed.): "The only complete manual written on the increasingly popular telescopic sight. Fully describes all makes and types of scopes in common use today. Tells just what rifles they are adapted to and how they should be attached. All about the modern, big game hunting scope; also varmint, small-game and target scopes when used with target or sporting rifles. A most authoritative and practical book from an author who has had 38 years of actual use and experience with the telescopic sight. 140 (sic) text pages and plates. Profuse with small illustrations."

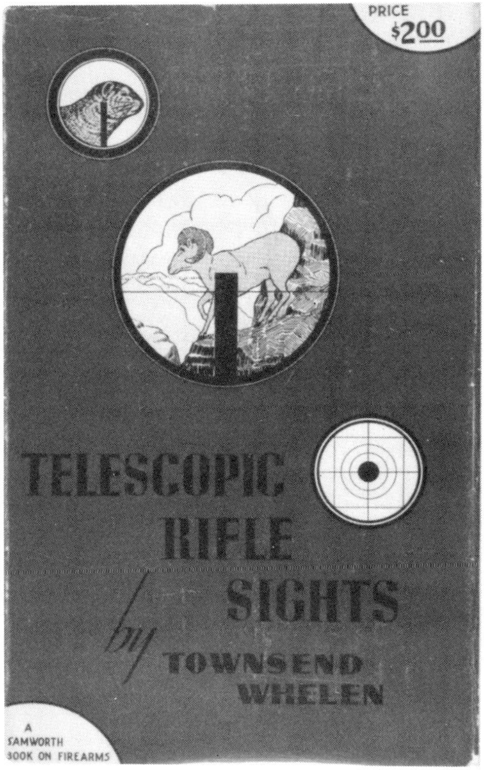

40) Telescopic Rifle Sights 1944 Edition
B. R. Smith collection

SUBSEQUENT PRINTINGS AND ISSUES:

c1936 Thomas G. Samworth Onslow County, NC. 1936 ad page date. 12mo. Case is black cowhide grain leatherette embossed cloth. Dust jacket is identical to that of the first impression. It is speculated that this was a later issue distributed shortly after the first was introduced and sold out; it is one possible explanation for the lack of a change in dates, and the subtle change in binding cloth..

c1936 Thomas G. Samworth Onslow County, NC. 1936 ad page date. 12mo. Case is a blue-black smooth cloth. Gilt lettering on front and spine. All edges dark blue ink-washed. Dust jacket is illustrated like that above, but jacket ads are different; four titles appear on each flap instead of two as in those above. "Samworth Books" cartridge case headstamp is printed at the top of the ads on the rear panel. Post-1936 titles are advertised.

SECOND EDITION:

c1944 Thomas G. Samworth Plantersville, SC. 1944 ad page date. 12mo.

CASE: Blue-black woven silk embossed stiff cloth. Gilt lettering on front and spine. All edges dark blue ink-washed. Dust jacket illustrated like that of the first edition, only "A Samworth Book on Firearms" appears in the semicircle on the lower left front corner. Dust jacket ads are different; two titles appear on each of the flaps, and the address is Plantersville.

CONTENTS: 199 text pp. Title page states: "Second Edition Revised and Enlarged." Eight chapters incl: General Principles; Target Scopes; Small Game and Varmint Scopes; Big Game Hunting Scopes; Snipers Scopes; Junior Scopes; Mounting Scopes on Rifles; Optics of the Scope. Two-page directory of suppliers follows.

Frontis. of the author posing in the prone position behind a scoped rifle. Illustrated with B&W plates, with sketches and line drawings by the author.

SAMWORTH PROMOTIONAL (1944 Ed.): "For more than a generation, Colonel Townsend Whelen has been one of our best qualified writers on firearms and all the appliances pertaining to them. He looked through his first rifle telescope (a Sidle) back in 1899. During the ensuing 46 years he has looked through hundreds of other telescopes--scopes of every make, size, type and power. He has killed many a head of big game and thousands of varmints with scope sighted rifles, also rolled up a few winning scores in matches where the 'glass eye' was permitted.

"This lifetime of experience with scopes has all been written into his *Telescopic Rifle Sights*, which is the only complete manual ever published on this increasingly popular type of sight. It recently went into a second and greatly enlarged edition. Herein are fully described and illustrated all makes and types of scopes now in common use. He tells you just what rifles these scopes are adapted to, just how they should be attached to those rifles, and just what the combination will be good for when completed. All about the modern big game hunting scope; also varmint, small-game, junior, sniping and target scopes when fitted to target or sporting rifles. A most authoritative and practical book of 199 text pages. Profuse with special plates and small sketches. A technical work highly necessary to every rifleshot. Any prospective scope purchaser will save money by first reading this manual."

SUBSEQUENT PRINTINGS AND ISSUES, 2ND. EDITION: None known.

NOTES: As stated earlier, Samworth did not go in for stated second editions of his books; only three titles ever had second editions, this being one of them. There is some confusion about this title due to the fact that the second edition was not widely advertised as being revised and enlarged. It appeared in post-1944 ad pages as *Telescopic Rifle Sights* (1944). As can be seen from the second edition promotional ad copy above, Samworth did call attention to the additional material included, but without fanfare. The book itself has "Second Edition--Revised and Enlarged" on the title page. Therefore, serious Samworth enthusiasts will probably want both editions represented in their collections. The second edition is perhaps one-eighth of an inch thicker than the first. Only one impression of the second edition has been documented, but it is probable that later issues exist due to the popularity of the title.

As with all "Small Sams", the first impression of this title came with the fragile thin pebble grain covers, which should not be subjected to unnecessary exercising. The later printing of the first edition has a handsome fine weave cloth binding which, although thin, holds up very well. The case on the second edition has thick boards which are quite robust.

Attention should be called to the difference in dust jackets between the first and second editions. Besides the ads being different, the front panel of the dust jackets are identical with the exception of the white semicircle at the lower left corner. The phrase "A Samworth Publication" appears in the semicircle on the first edition jacket, but that of the second edition is printed with "A Samworth Book on Firearms."

As a sidelight, for anyone interested in pre-WW II scopes and mounts, this little manual reads like a who's who: Belding & Mull, Fecker, Hensoldt, Lyman, Malcolm, Noske, Redfield, Unertl, Weaver, Wollensak, Zeiss, etc. Whelen enthusiasts will also have an interest here as well. Examples of each edition can be found, and there appears to be strong, steady demand for them. Many dealers do not recognize the fact that two editions exist, nor do they seem to differentiate with respect to pricing.

The copyright page of the first edition lists other books by the author, only one of which is a Samworth. The publisher was not in the habit of mentioning other titles written by those authors in his stable.

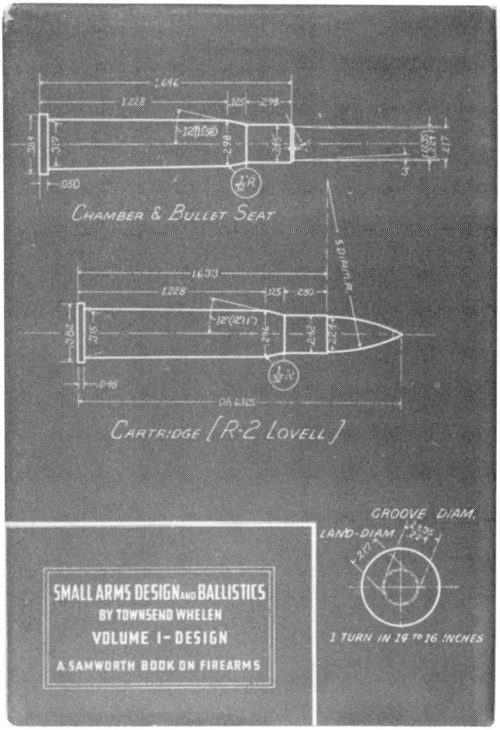

41) Small Arms Design and Ballistics
Vol. I - Design
B. R. Smith collection

41) and 42) WHELEN, Townsend

SMALL ARMS DESIGN AND BALLISTICS

41) Volume I: SMALL ARMS DESIGN

c1945 Thomas G. Samworth Plantersville, SC. Dec. 1944 ad page date. Crown 8vo.

CASE: Medium-blue heavily filled buckram. Gilt lettering overprinted on light gray title blocks on front and spine. "Samworth" in gilt on a light grey block above tail cap. Dust jacket illustrated in blueprint motif with chamber and cartridge dimensioned drawings of the R-2 Lovell cartridge on the front, two bullets on the spine, and ads on the back panels. Book title appears in a stylized blueprint title block at the lower left corner of the front panel.

CONTENTS: 352 pp. incl. index. Foreword and twelve chapters incl: Small Arms in General; The Rifled Bore; Barrels in General; Manually Operated Breech Actions; Semi-Automatic Breech Actions; Stocks and Sights; Ammunition-Introductory; Ammunition-Primers; Ammunition-Cases and Shells; Ammunition-Propellant Powders; Ammunition-Bullets; Ammunition-The Loaded Cartridge. A brief commentary "Advantages of Superior Arms" follows.

Frontis. of the author shooting a rifle at his bench. Illustrated with B&W halftones, and sketches by Roger Marsh.

SAMWORTH PROMOTIONAL: "This book covers the entire field of modern small arms and ammunition, with special stress on design, construction and operation. The author, Col. Whelen, an outstanding rifleman and small arms authority and one of the U.S.

Army's leading ordnance experts, has made the study and use of firearms his hobby and profession for more than half a century.

"Although replete with technical notes and data, Volume I is devoid of higher mathematics. The vast subject is covered in a clear and masterful manner by the writer, easily understood by the average man, thereby affording an opportunity for the layman to grasp and apply solutions to his problems, which, unfortunately, all too often is not the case where technical or semi-technical works are concerned. This is a book the beginner will easily understand and is a splendid work to start off with.

"Volume I not only thoroughly covers the construction, design, and operation of small arms--including rifles, pistols, revolvers, shotguns, and submachine guns--but gives minute techincal details of cartridges and ammunition components. This concise, authentic information regarding design, manufacture, functioning and use of small arms and ammunition constitutes a most liberal education for any shooter and the entire subject is presented in a manner that he can absorb and utilize in a practical way in his everyday shooting problems."

SUBSEQUENT PRINTINGS AND ISSUES:

c1945 Thomas G. Samworth Georgetown, SC. 1945 ad page date. Crown 8vo. Case is similar to that of the first impression; overprinted lettering panels are a medium blue instead of light gray. Text paper is a creamier shade. Dust jacket has the blueprint-like illustrations.

c1945 Thomas G. Samworth Plantersville, SC. Mar. 1946 ad page date. Crown 8vo. Second issue of the first impression. Case is medium-blue heavily filled buckram. Gilt lettering overprinted on light gray title blocks on front and spine. "Samworth" in gilt on a light grey block above tail cap. Dust jacket illustrated in blueprint motif with chamber and cartridge dimensioned drawings of the R-2 Lovell cartridge on the front, two bullets on the spine, and ads on the back panels. Book title appears in a stylized blueprint title block at the lower left corner of the front panel.

SUBSEQUENT EDITONS: Various Stackpole impressions, so marked on spine, title page and ad page.

NOTES on Volume I follow the description of Volume II.

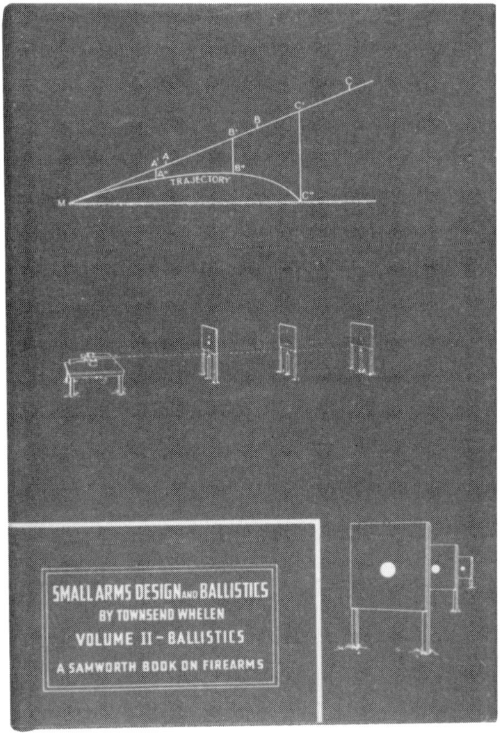

42) Small Arms Design and Ballistics
Vol. II - Ballistics
B. R. Smith collection

42) Volume II: SMALL ARMS BALLISTICS

c 1946 Thomas G. Samworth Georgetown, SC. 1946 ad page date. 1946 title page date. Crown 8vo.

CASE: Medium-blue filled buckram. Gilt lettering overprinted on light gray title blocks on front and spine. "Samworth" in gilt on a light gray block above tail cap. Gray bullseye with the numeral "2" in the center is in the middle of the spine. Dust jacket illustrated in blueprint motif with a geometric plot of a projectile trajectory in relation to line of bore; a benchrest and three colinear target stands appear below. Book title appears in a stylized blueprint title block at the lower left corner of the front panel. Two sectioned cartridges on spine depicting the progression of primer ignition through low and high density powder charges.

CONTENTS: 314 pp. incl. index. Introduction and thirteen chapters incl: Interior Ballistics; Determination of Pressure and Velocity; Recoil, Jump, and Vibration; Exterior Ballistics; Trajectory--A Study; Determination of Trajectory; Wind Deflection; Wounding Effect and Killing Power; Testing Ranges, Rests, and Records; Shotgun Ballistics; Care and Storage of Arms and Ammunition; Hand Loading Ammunition; Old Records, Experiences and Opinions. Appendix follows.

Frontis. of the author and J. Bushnell Smith at a shooting bench. Illustrated with halftone plates and line drawings. Five loose, folded sheets containing the ten E. I. Du Pont de Nemours Ballistic Charts by engineers Coxe and Bugless (sic) are laid-in.

SAMWORTH PROMOTIONAL: "In Volume II Colonel Whelen thoroughly covers important and hitherto vague subjects of Interior and Exterior Ballistics. This work clearly and graphically depicts just what happens from the time the trigger is let off until the bullet strikes the target.

"Volume II differs from other textbooks on small arms ballistics in that the writer treats the subject in plain language that the average reader can understand, rather than in cold erudite terms and a maze of advanced mathematics which oft-times puzzle and bewilder even learned scholars and scientists.

"This book, complete in itself, presents extensive explanations and studies relative to pressure, velocity, combustion, recoil, jump, vibration, flight of the bullet, air resistance, force of gravity, drift, yaw, trajectory, accuracy, killing power, testing methods, bench and machine rests, care, cleaning and storage of materials and the necessary methods of computing and keeping proper records and scorebooks.

"The entire subject is approached from the practical, rather than the scientific angle, and the text will be found of particular benefit to the average shooter and hunter. The set comprises a basic foundation of ballistic knowledge for arms students, ordnance technicians, sportsmen and firearms lovers in general."

SUBSEQUENT PRINTINGS AND ISSUES:

c1945 Thomas G. Samworth Georgetown, SC. 1955 ad page date. Crown 8vo. Case is similar to that of the first impression; overprinted lettering panels are a medium blue instead of light gray. Dust jacket has the blueprint-like illustrations.

SUBSEQUENT EDITIONS: At least two Stackpole impressions exist, so marked on spine, title page and ad page.

Advertised as being available in 1990; facsimile reprinted in deluxe quarter-leather binding by Wolfe Publishing Co. in the Wolfe Library Classics series.

NOTES: Considered by many to be among the rarest Samworths; the person who seeks these titles tends to keep them. For content they are nearly without equal, especially Volume II. Stackpole reprinted both volumes in the '60s to satisfy demand. Judging by present activity on the used and rare market, it is ripe for reprint again, as the Wolfe Publishing reprint edition attests.

Volume II is sought after moreso than Volume I, and therefore is the most scarce. The first volume is occasionally seen alone; sets are unforunately broken up to meet demand for Volume II. Some sets have been seen with mixed Samworth imprints, as well as Samworth and Stackpole editions, and the serious collector should be aware of this. It is recommended that the title page, ad page and spine on both volumes be examined before purchase to identify the impressions.

The first impression of Volume II had the numeral "2" in a gray circle printed in the center of the spine. There has been some confusion concerning this adornment since the first impression of Volume I lacks any such corresponding decoration. Volume II was published over a year after the appearance of Volume I, and for some unknown reason it was decided to mark the spine with a number indicating the second volume. Some confusion exists regarding Volume II; it is widely believed by many that the second volume was sold only together in a set with Volume I, and this is not the case. Additionally, it should be stressed that the correct publication date for Volume II is 1946.

Collectors often ask how the first issue of Volume I can have an ad page date of December 1944 and a copyright date of 1945. This stems from the fact that the book was typeset late in 1944, and the work was not published until the next year when the copyright was registered.

The 1955 Samworth impressions have title panels printed in a medium blue rather than a gray, with a gilt lettering overprint. The bindings of both volumes are a dark medium blue, not quite navy.

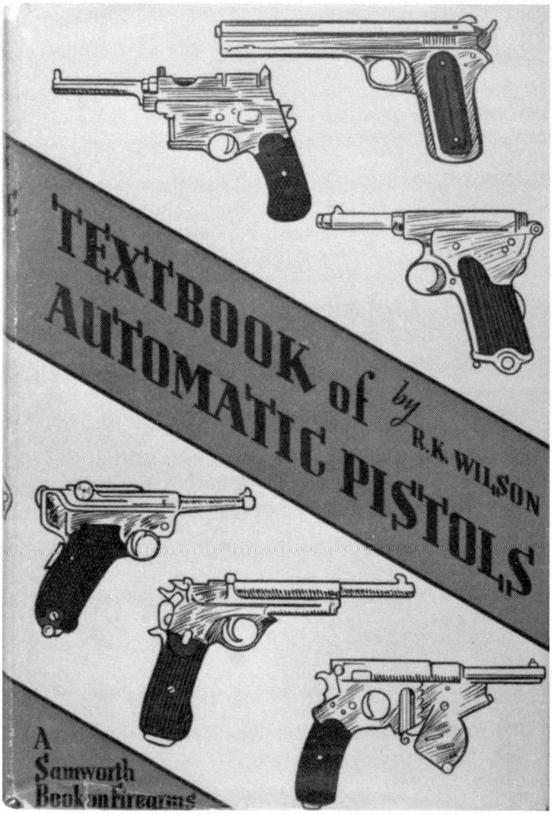

43) Textbook of Automatic Pistols
B. R. Smith collection

43) WILSON, R. K.

TEXTBOOK OF AUTOMATIC PISTOLS

Being a treatise on the history, development and functioning of the modern military self-loading pistol--its special ammunition--and their evolvement into the sub-machine gun--together with a supplementing chapter on the light machine gun, 1884-1935.

c1943 Thomas G. Samworth Plantersville, SC. 1943 ad page date. Royal 8vo.

CASE: Maroon coarse weave filled cloth. Gilt lettering overprinted on kelly-green panels on front and spine. "Samworth" in gilt above tailcap. Dust jacket illustrated with drawings of automatic pistols; title is printed in a diagonal band across the center of the front panel. Ads appear on the back and both flaps.

CONTENTS: 349 pp. incl. index. Preface and twelve chapters incl: Historical Survey: The Development of the Automatic Principle; Principles of Operation; Early Self-Loading Pistols; Mauser and Mauser-Type Pistols; Military Self-Loading Pistols (1); Military Self-Loading Pistols (2); The U. S. Army Trials of 1907 and the U. S. Army Model Colt; Intermediate and Police Self-Loading Pistols; Some Pocket Self-Loading Pistols; Calibre .380; Automatic Pistol Ammunition and Ballistics; Pistol-Cartridge Automatics: Sub-Machine Guns and Light Automatics for Low-Power Ammunition; Light Automatics; A Short Out-line of Development.

Frontis. of the 8mm Schonberger pistol from 1892. Illustrated with plates showing 55 pistols and machine guns, and line drawings of cartridges in the text.

SAMWORTH PROMOTIONAL: "Admittedly the finest and most thorough book yet written on its subject. The author, an officer in the Royal Artillery of the British Army, is among the foremost authorities on automatic autoloading weapons: this book is the product of his knowledge and experience. It progresses from a basic treatment of automatic breech mechanisms to the specific weapons themselves, beginning with the Schonberger and continuing through the most modern Le Francais. Such little-known weapons as the Gabbet-Fairfax 'Mars' and the original .455 Webley 1904 are here covered fully for the first time. An entire chapter is devoted to the various models of the Military Mauser and its imitations, while another section covers the Lugers, Steyr-Hahn, and other heavy military pistols. Such interesting American arms as the Grant-Hammond and the experimental Savage .45 are fully dealt with, while in another portion of the book is found brief mention of a number of pocket pistols.

"One feature of particular interest is the section of this work devoted to ammunition adapted to automatic pistols--every cartridge designed specifically for self-loading pistols is described and full data on it is given. Also especially valuable to designers are the lists of patents relating to small arms.

"In spite of its title, however, this book is not wholly confined to automatic pistols. The person interested in sub-machineguns will find here complete information on such 'pistol cartridge automatics' as the Villar Perosa, the Mauser, Bergmann Musquete, and the superb Neuhausen gun; nor is the Thompson SMG ignored. The author has also included an extensive chapter on the earlier light automatic shoulder rifles, from the Madsen to the Bren, Darne, and Vickers-Berthier guns.

"This is without doubt one of the most extensive and finest books on automatic weapons in general and automatic pistols in particular which has ever been written. It is clearly and profusely illustrated with both text cuts and plates, most of the latter showing two different firearms. 349 pages, 55 plates, indexed."

SUBSEQUENT PRINTINGS AND ISSUES:

c1943 Thomas G. Samworth Plantersville, SC. 1944 ad page date. 8vo. Later issue. Case and dust jacket are identical to that of the first impression.

c1943 Thomas G. Samworth Georgetown, SC. 1950 ad page date. 8vo. Case is a dark charcoal-gray coarse buckram. Dust jacket is also similar but ads on back flaps are different.

SUBSEQUENT EDITIONS:

No known later editions. A Stackpole editon may exist, but a copy has not been examined to date.

Advertised as being available in 1991; facsimile reprinted in deluxe quarter-leather-bound format by Wolfe Publishing Co. in the Wolfe Library Classics series.

NOTES: Long a definitive reference, the material is somewhat dated by present standards; it contains nothing on post-WW II pistol developments. However, there is strong demand for the title from enthusiasts of early autoloading pistols, rifles, and sub-machine guns, who consider it a "bible" of sorts. Relatively scarce as most owners tend to keep the work; there are few other references that describe in detail the many obscure and generally unheard-of weapons that Wilson's does.

The plate illustrations of these pistols are very interesting to the collector or student of these arms because few other references contain them, even if the pistols are mentioned in the text.

There is no better book available on early autoloading arms.

FIREARM DESIGN AND ASSEMBLY BOOKLETS

The next four entries, *The Inletting of Gunstock Blanks*, *The Shaping of Inletted Blanks*, *Finishing the Gunstock*, and *Checkering and Carving the Gunstock*, Nos. 43 through 46 respectively, document the large-format wrappers-bound "booklets" on stockmaking and checkering published by Samworth beginning in 1941. Although many titles were planned and announced, only these four were actually published.

The first three titles in the series were written by the renowned stockmaker Alvin Linden. He had an arrangement with Samworth to provide three additional titles, *Checking and Checking Patterns*, with pattern sheets for the Mauser rifle; *Metal Parts and Operations*, with pattern sheets for the Krag rifle; and *Repairs and Refinishing--The Design of Gunstocks*. These were not completed due to Linden's failing health and subsequent death in 1946. Each of the published titles advertises one or more of the three uncompleted works, thus giving rise to some considerable confusion among serious book collectors and dealers alike.

In a departure from the format used for preceding hardbound entries, this section on the *Firearms Design and Assembly* booklets begins with a Samworth promotional for the entire series. The entries for the individual titles follow, with complete descriptions of all respective details.

SAMWORTH SERIES PROMOTIONAL:

FIREARM DESIGN AND ASSEMBLY

"These detailed and authoritative works are offered with a view of presenting the maximum amount of correct and practicable gunstocking instruction in the most applicable manner--by means of accurate full-scale patterns and detail drawings of modern, custom built gunstocks; supplemented by a string of sketches and photographs showing various details, operation procedure (sic) and mechanical set-ups necessary to carry out the work.

"These booklets have been written and published for the benefit of the individual who is actively engaged in gunsmithing, or who intends to take it up in a serious way; they are not works which give merely a general knowledge of the art as a whole but are of a highly detailed and comprehensive nature. They tell all of the how and why. The series is being got up in a manner best adapted to practical application and use, yet in a format permitting the most economical price possible.

"Throughout this series, the various stages of gunwork have been treated separately, in individual booklets, each of which is distinct yet complete in its scope and treatment. Texts run from 30,000 to 50,000 words to the booklet, with many specific and original drawings and photographs used to illustrate the various steps, operations and layouts. There is no duplication of any nature throughout this series, each booklet being an original work and taking up where the preceding title left off. Hence, it is not necessary to purchase the entire series in order to obtain information or guidance on any particular stage of gunsmithing practice, the pertinent title will be found to contain full material and instruction on that phase of the work.

"The most unique (and original) feature of this series is the two large "wall chart" pattern sheets included with every booklet, each of which has been composed to accompany that particular booklet subject. Each of these pattern sheets show two deluxe examples of custom-built sporting or target rifles in all their details; four full-scale patterns with each booklet. Size of the sheet is 26" X 40", enabling everything to be drawn to full-scale; any desired pattern or dimension can be traced or scaled-off directly from the sheet. With these accurately drawn patterns to follow, you will have no trouble in getting your first gunstock correct in every particular.

"Each set of patterns differs radically from all the others, each has its own distinctive details and form, yet it is an easy matter to apply or transfer the more desirable features of a stock to some other assembly you like better, or to a rifle of another model. These 'life size' pattern sheets must be seen and closely studied before their full value and application can be realized, as their possibilities are almost unlimited. In fact, these Linden pattern sheets represent to the stockmaker what the conventional dress pattern means to the home dressmaker. In particular, they can be a great value to the professional gunsmith in arriving at a clear understanding with a prospective customer regarding the style and exact dimensions of the rifle under discussion or about to be built.

"The author of this series was a 'full time' professional gunsmith of national reputation and standing--the late Alvin Linden, of Bryant, Wisconsin. Linden had estblished an outstanding name for himself during the past two decades as a gunstocker whose work was unsurpassed, and he was also renowned as an ace designer of modern gunstocks for every purpose and need. His output could not begin to meet the innumerable demands for a 'Linden stock' and his advice was eagerly sought on all questions regarding rifles and shotguns. This series of works reflected his artistic temperament and commonsense opinions. He practiced his profession 'back in the sticks' of Langlade County and this backwoods location greatly enhanced his extremely practical viewpoint on any and all matters relating to the construction and use of firearms. Best of all, he wrote in shirt-and-drawers English, readily understandable to all and with plenty of he-man peeking out now and then. You will enjoy his many localisms and the homely humor interspiced amongst his clear explanations and teachings.

"Text, photographs, drawings and patterns have all been done either by Linden or under his personal supervision--made right there in the Bryant gunshop--insuring a work complete and correct in every particular. The character of the man, as well as his ability, is clearly shown in these works and we are proud to offer them as Samworth Books on Firearms."

44) The Inletting of Gunstock Blanks
Fred Talkington collection

44) LINDEN, Alvin

THE INLETTING OF GUNSTOCK BLANKS
as Applied to Modifications of the Springfield Rifle. Being the first booklet in a series devoted to the Design and Assembly of Custom-built Firearms.

c1941 Thomas G. Samworth Plantersville, SC. 1941 ad page date. Small 4to.

COVER: Paper wrappers; cream stock with horizontal green bands and black lettering. Covers are printed with ads on all panels but the front. No dust jacket.

CONTENTS: 38 pp. Undivided text printed on coated stock details the inletting of gunstocks from the selection of the blank to finishing the barrel channel for bedding.

Frontis. of the author at work. Illustrated with halftones and sketches. Folded pattern sheet Nos. 1 and 3 laid in.

SAMWORTH PROMOTIONAL: "Inletting the Stock Blank is a subject which, in the past, has invariably been covered in one brief chapter, but here is an entire book on this operation alone. Forty large pages containing better than 30,000 words, with 41 operation sketches and four halftone illustrations.

"Here is the real step-by-step procedure of inletting the rough blank for a Springfield barrel and action. Several exceedingly practical inletting tools are told of for the first time, and their proper use shown. The instructional text is interspiced with discussions on the correct principles of Bedding, Layout of Grain, Selection of Wood, Proper Seasoning, and other equally important subjects. This is not one of those 'touch lightly' effusions wherein the author skips vaguely over many pertinent phases of inletting with a few brief paragraphs, due to a lack of actual experience on such subjects. Linden has inletted hundreds of gunstock blanks and he here tells you ALL about it--or at least 30,000 words-worth on the matter of inletting and nothing else. The principles, operations and procedure given are applicable to any model of bolt-action rifle.

"The two accompanying Pattern Sheets have been wrapped around the subject of inletting and they show four examples of custom-built Springfields, three of which are hunting models of varying weights, while the fourth is a heavy-barreled match rifle of superior possibilities. The patterns clearly illustrated several heretofore obscure points relative to correct inletting, and a gander or two taken at thcsc full-scale drawings while on that job will enable the worker to fit and bed his barrel and action in accordance with modern theory and practice."

SUBSEQUENT PRINTINGS AND ISSUES:

c1941 Thomas G. Samworth Plantersville, SC. 1943 ad page date. Small 4to.

c1941 Thomas G. Samworth Georgetown, SC. 1949 ad page date. Small 4to.

c1941 Thomas G. Samworth Georgetown, SC. 1954 ad page date. Small 4to.

45) The Shaping of Inletted Blanks
Fred Talkington collection

45) LINDEN, Alvin

THE SHAPING OF INLETTED BLANKS

with Notes on Some Alterations to the Model 70 Winchester Rifle. Being the second Booklet of a series devoted to the Design and Assembly of Custom-Built Firearms.

c1941 Thomas G. Samworth Plantersville, SC. 1941 ad page date. Small 4to.

COVER: Paper wrappers; tan stock with horizontal red bands and black lettering. Covers are printed with ads on all panels except the front. No dust jacket.

CONTENTS: 54 pp. Undivided text printed on coated stock details the shaping of gunstocks, sizing for proper fit, and the installation of hardware fittings.
Frontis. of a custom-stocked Winchester M70 in .300 Magnum. Illustrated with halftones and sketches. Folded pattern sheets Nos. 5 and 6 laid in.

SAMWORTH PROMOTIONAL: "The Shaping of Gunstocks (sic) comprises some 45,000 words, contained in a 54 page booklet, with a great many layout sketches, the necessary operation and sequence drawings and, of course, plenty of halftone illustrations to cover the main whacks and trimmings.

"In the past, a great many gun authorities have always overestimated the difficulties to be met up with while the shaping of the stock and forearm is being carried out, and in previous literature much stress has been laid on in the necessity of considerable inborn talent or inherent skill in order to produce satisfactory stocking lines; readers have been constantly impressed with the need for bringing into play their 'hidden talent', 'artistic expression', etc. Well, the plain fact is that the average chap, whose only means of expression is apt to be profane English, between that and this No. 2 Linden manual can now turn out a gunstock of correct and pleasing lines ON HIS FIRST ATTEMPT. He MIGHT even leave out the cussin' and still have his stock fit, look and be right.

"Heretofore, the stumbling block to most beginners has been the proper layout of the basic points of the gunstock, not the working-down to its final lines. Layout, like most of the things in this world of ours, is exceedingly simple once you understand the correct way of going about it, and Linden has herein gone to considerable extent and wordage to show how comparatively easy it is to CORRECTLY lay-down the stipulated or desired lines and shape on the inletted blank BEFORE working away any of the wood.

"In addition, all the necessary details as to the trimming off to the proper outlines is included, down to the last small dimension and fixture. The making of several shaping tools is described in detail, also illustrated; make them right there in your own workshop, from scrap stock lying around.

"Everything is written in shirt-and-drawers English, clearly understandable to the ordinary dub, yet extremely instructive to advanced workers also. Here is the necessary 'know how' that will enable YOU to get that first gunstock shaped RIGHT and at the same time have it FIT the specifications you or anyone else may lay down.

"Pattern Sheets Nos. 5 and 6, which accompany this booklet, show four examples of the Model 70 Winchester rifle; one a splendid full length a-la-Mannlicher stocked affair; and one a very practical straight-gripped job that will take the eye of many big-game hunters. These sheets also illustrate layout lines and methods relative to proper positioning of the pistol grip and the comb nose, two important points which are often difficult for both customer and stocker to get right . . . or get in the right place."

SUBSEQUENT PRINTINGS AND ISSUES:

c1941 Thomas G. Samworth Plantersville, SC. 1943 ad page date. Small 4to.

c1941 Thomas G. Samworth Georgetown, SC. 1949 ad page date. Small 4to.

c1941 Thomas G. Samworth Georgetown, SC. 1951 ad page date. Small 4to.

c1941 Thomas G. Samworth Georgetown, SC. 1954 ad page date. Small 4to.

46) The Finishing of Gunstocks
Fred Talkington collection

46) LINDEN, Alvin

THE FINISHING OF GUNSTOCKS

and Notes on the Conversion of the 1917 Enfield Rifle. Being the third Booklet in a series devoted to the Design and Assembly of Custom-Built Firearms.

c1941 Thomas G. Samworth Plantersville, SC. 1941 ad page date. Small 4to.

COVER: Paper wrappers; yellow stock with horizontal green bands and black lettering. Covers are printed with ads on all panels except the front. No dust jacket.

CONTENTS: 46 pp. Undivided text printed on coated stock details the preparation of gunstocks for the application of various finishes; the second half of the text concerns detailed modifications required to custom sporterize the M1917 Enfield rifle.

Frontis. of a custom-stocked sporterized M1917 Enfield. Illustrated with halftones and sketches. Folded pattern sheets Nos. 2 and 4 laid in.

SAMWORTH PROMOTIONAL: "The Finishing of Gunstocks consists of some 38,000 words, contained in 48 pages, with 15 special illustrations of differenly finished gunstocks; plus some detail sketches of various modifications to the Enfield barrel, sights and action.

"The entire subject of stock finishes and finishing has been gone into in minute detail, and for the first time the underlying principles of process and manner of application have been discussed--and NOT merely a few possibly-useable formulas given; which may or may not work in your particular case. Once you have soaked up the information contained in this booklet, you will be able to compound your own workable stock finish; and will know how to apply it after you have also studied the particular piece of wood to which it must be applied. All the commonly used finishing oils and varnishes are analyzed and discussed in detail from the gun crank's viewpoint, with a few new possibilities pointed out.

"Also, and for the first time, the reader is initiated into the mysteries of 'Suigi' finishing a gunstock; which is nothing more or less than graining it with a blowtorch--as any half-soused house painter will point out. Hot Dawg! but you boys will go wild when you realize the possibilities of this stunt; which seems to have been lost in the shuffle during the past several generations, although our colonial ancestors did right well with it in figuring and bringing out the grain on their flintlock stocks. Well, here it is, a little late possibly, but here again at last.

"Then, there are some 16 pages crammed with dope on the various alterations and operations necessitated when remodeling the Model 1917 Enfield into a sporter; all being specialized data pertinent to this rifle and its peculiar problems of magazine fit and three-point bedding of the barrel and action.

"The Pattern Sheets have been composed to accompany the booklet subjects. They show deluxe examples of four custom-built Enfields, whereon the utmost in grain and figuration have been brought out with Suigi treatment. Three are Enfield sporting models of advanced design; the fourth is a heavy barrel, single shot version of a scope-sighted varmint outfit in .220 Swift caliber, a peculiarly practical and unique assembly."

SUBSEQUENT PRINTINGS AND ISSUES:

c1941 Thomas G. Samworth Plantersville, SC. 1943 ad page date. Small 4to.

c1941 Thomas G. Samworth Georgetown, SC. 1948 ad page date. Small 4to.

SUBSEQUENT EDITIONS:

Impressions of all three Linden FD&A Booklets exist with Samworth and Stackpole markings. The latter were Georgetown impressions from Samworth's remaining stock that were sold to Stackpole in the late 1950s, and are seen with paste-on price stickers and sometimes paste-over labels with the Stackpole address. Some of the Stackpole-marked impressions have such identification printed directly on the wrapper covers.

All three Linden booklets were consolidated in a hardbound volume and published by Stackpole under the title *Firearms Design and Assembly--Restocking the Rifle* which is no longer in print and highly sought after in its own right. It included all pattern sheets for all three Linden booklets in a pocket attached to the inside back cover.

NOTES ON THE LINDEN BOOKLETS: Serious enthusiasts are cautioned to examine copies for purchase to make certain that the proper two pattern sheets for that volume are laid-in; only when they are included should a title be considered complete. Also, the text is saddle-stitched to the cover with staples magazine-style, and these titles should not be opened unless absolutely necessary. For reading or reference worn copies of individual booklets or a Stackpole three-in-one volume noted above should be obtained, as the staples easily pull out of the cover fold. Text paper is a high-quality coated stock which seems to hold up well with time, as evidenced by the copies examined.

47) Checkering and Carving of Gunstocks
Robert J. Wos collection

47) KENNEDY, Monty

CHECKERING AND CARVING OF GUNSTOCKS
The fourth Booklet of a series on the Design and Assembly of Custom-Built Firearms.

c1952 Thomas G. Samworth Georgetown, SC. 1952 ad page date. Small 4to.

COVER: Pale blue stiff pasteboard with black spine and red horizontal bands on the front. "Samworth Book" cartridge case head on back.

CONTENTS: 244 pp. plus ads. Unchaptered text is printed on heavy coated stock.

Frontis. of the late Alvin Linden at work. Illustrated with halftones, line drawings and sketches. Material includes layout of checkering patterns, use of checkering tools, and examples of others' work. Pattern sheets are bound-in unlike those of the three preceeding FD&A booklets which are laid-in.

SAMWORTH PROMOTIONAL: "Gunstock ornamentation has been a subject of compelling personal interest to most owners of firearms for the past five centuries and has ranged from the practice of hammering-in lines and whorls of brass nails on up to the meticulous inlaying of mother-of-pearl and semi-precious stones into the wooden stock. Such decorative tastes have leveled off in the past 100 years and now are confined mainly to the checkering and carving of the gunstock with a minimum of inlays.

"During the past decade American gunmakers have been hard put to meet the increasing demand for more artistic ornamentation to grips and forearms, yet, at the same time, meet the necessary requirements prescribed by usage in hunting fields and on target ranges.

"Checkering and Carving of Gunstocks meets the demand for a treatise covering this art of gunstock ornamentation. It is an extensive and specialized work of 244 text pages and some 300 technical illustrations that covers its subject fully from both utilitarian and decorative standpoints. The author, Monty Kennedy, is a top flight professional gunstocker with several years of steady work at the checkering cradle to back him up, and he tells herein all he knows.

"In addition to the author's comprehensive instruction, this book gives the methods, procedures, tools and patterns of such top-ranking craftsmen as Tom Shelhamer, Leonard Mews, John Hearn, Leighton Baker, Keith Stegall, Hal Hartley, Roy Dunlap and many others. Each describes at length his distinctive checkering methods and illustrates his personal choices of actual patterns.

"There are some 75 of these checkering patterns shown--each in full size with basic starting lines located--all ready to be used as a template of sorts for transfer to your own gunstock. Notes and comment as to its application by the beginner accompany each pattern. Patterns range from easy ones to some that will take many, many jobs before experience enough has been acquired to start in on them.

"The past few years have seen an increasing demand for carved gunstocks, so to help meet this growing trend we present, in full size, some 40 applicable and highly attractive carving patterns--forearm, grip and buttside--of big game, game bird, animal, and leaf-and-seed designs. Nothing conventional here; these are definite, easily recognizable patterns from nature's storehouse, highly original in conception and done by leading American artists. The beauty and appeal of these splendid patterns must be seen to be appreciated and this feature will set a new milestone in custom decoration by American gunmakers. Accompanying these patterns is the essential basic instruction on carving and suggestions as to the application of these particular designs. Monty Kennedy's great work will prove of equal value to the experienced gunmaker with years of practical experience to his credit, as well as to the man or boy out in the hills with his first standard grade of a gun and a bit concerned about that smooth, unfinished look to its grip and forearm. Either of these extremes will find something in *Checkering and Carving of Gunstocks* that can be applied to the gun-in-hand at the particular moment."

SUBSEQUENT PRINTINGS AND ISSUES:

c 1952 Thomas G. Samworth Georgetown, SC. 1953 ad page date. Small 4to.

c 1952 Thomas G. Samworth Georgetown, SC. 1954 ad page date. Small 4to.

Covers on these printings are identical to that of the first impression with the exception of the spine tape seen on some copies, which is red instead of black. The ads at the end of the text are different as well.

SUBSEQUENT EDITIONS:

A Stackpole impression exists that is identical to the 1954 Samworth. Stackpole later brought out the same title in a true cloth hardbound edition, updated with many pages of additional material.

NOTES: This is the fourth and final Firearms Design and Assembly title in the series. It should be noted that this title is listed as No. 4 although the first three titles in the series advertised the title of No. 4 as *Checking and Checking Patterns*, with patterns for the Mauser and its adaptations. As stated earlier, this title was never published since its author, Alvin Linden, passed away before its completion. The Kennedy work took its place.

Checkering and Carving of Gunstocks differs from its series predecessors in that it is hardbound with pasteboard covers, and the text is at least four times thicker than the Lindens. Furthermore, the large, folded patterns are printed on uncoated stock and bound-

bound-in as well. Thus, the collector need not worry about missing pattern sheets with the Kennedy book as with the Lindens.

Most examples of this title display severe binding welts on the free endpapers and flyleaves. This was a binding fault and is not indicative of moisture damage.

Checkering and Carving of Gunstocks is a relatively common title that is often seen advertised in catalogs and on dealer's shelves. Condition tends to vary widely, depending on the amount of use to which the previous owner subjected it.

SATPCO TITLES NEVER PUBLISHED

The following list of titles were advertised by Samworth as "in preparation", "available soon", etc. In some cases the title was changed when the book was actually published, or the situation changed where the book was produced by another publisher. However, most of them never carried the Small Arms Technical Publishing Co. imprint.

The collector should note well the fact that early SATPCO sales literature and advertisements listed titles published by other firms. Samworth intended to function as a "full-service" firearms-related bookseller at the insistence of his associate, Townsend Whelen. Although the practice was dropped a few years after the firm began, this aspect of the business is one of the major sources of confusion and misinformation regarding the Small Arms Technical Publishing Co.

KEITH, Elmer

Varmint and Small Game Rifles Advertised but never published; Keith refers to this book in his two autobiographies as having been written along with the other two, and subsequently purchased by Samworth for $250. Samworth once made the comment that Keith's books had not been particularly successful for him, and the *VARMINT* manuscript took third place in his estimation. If the manuscript still exists, it would certainly be priceless.

CROSSMAN, E. C.

Target and Sporting Rifle Shooting Advertised as being "in preparation", this work was split into two published titles: *The Book of the Springfield* and *Military and Sporting Rifle Shooting*, being in the end too much material for any single volume.

THOMAS, Chauncey

Campfire Talks Not published by the Small Arms Technical Publishing Co. Advertised in early Samworth sales literature along with some titles from other publishers.

FITZGERALD, J. H.

Pistol Shooting Became *Shooting*, c1930 G. F. Book Co. Samworth worked closely with the author to produce a work that would appcal to the intended audience, the original manuscript falling rather short of the mark. After considerable effort on Samworth's part, a finished draft was returned for "proofing" to Fitzgerald who decided for unknown reasons to contact a different publisher and ultimately strike a deal with G. F. Book Co. The actual transpirings are outlined in a February 1, 1930 letter from Samworth to a Mr. Seth Wiard of the Northwestern University Scientific Crime Detection Laboratory. Mr. Wiard also contributed m+aterial to Fitzgerald's book.

WHELAN, Townsend

American Sporting Rifles Originally presented in *The American Rifle*, some of this material was incorporated into *The Hunting Rifle*, c1940 Stackpole & Heck.

SMITH, Ray M.

Pennsylvania Rifles and Riflemakers
Advertised repeatedly over a period of years, this work never materialized due to incompletion of the manuscript and differences of opinion between author and publisher as to editorial content. The compiler has on file quotations from Kingsport Press outlining

the details of the book and cost estimates for a 3000-copy press run, one dated 1947 and the other dated 1951. Associates of Samworth indicated that he was undecided as to whether a chapter on fakes should be included; the nature of such information was so complete and detailed that numerous collectors of these very expensive antique weapons would find they really did not have what they believed. Between author and publisher a substantial amount of evidence was obtained with respect to Kentucky rifles that had been "modified" by unscrupulous gunsmiths into "authentic" and thus more valuable Pennsylvania rifles. At any rate, the manuscript languished for years and mysteriously disappeared from Samworth's third floor office-study at Dirleton during an absence.

The dust jacket illustration was to be "The Rifle Frolic", one of the two Samworth Prints on Firearms painted by Gayle Hoskins. It has been confirmed that the text illustrations were commissioned and completed for this work by E. Stanley (Ned) Smith. They are currently in the possession of a surviving friend of Samworth's.

LANDIS, C. S.

Book of the .22 This was the prepublication title of *Twenty-Two Caliber Varmint Rifles.*

BAKER, L. L.

Single-Shot Rifle Conversions The manuscript was not completed by the time Samworth retired, and is still in the author's possession. In a conversation with the compiler, Mr. Baker related that the book was about half finished when single-shot rifles became recognized as valuable collectors' pieces in their own right, thus driving up the price of what had been inexpensive, well-made actions for use in building up custom target and varmint rifles.

LIND, H. A.

Japanese Rifle Conversions The manuscript was not completed by the time Samworth retired. There was also the post-war public conception of things Japanese being decidedly inferior to anything else, despite extensive testing on the Japanese military rifle actions which clearly demonstrated superior strength .

TRIEBEL, Wilhelm R.

German Gunstock Ornamentation The manuscript was essentially completed and illustration drawings were commissioned but the book was not published due to Samworth's retirement.

LINDEN, Alvin

Metal Parts and Operations Was to be one of the *Firearm Design and Assembly* series of booklets. Linden passed away in 1946 before the manuscript was completed.

LINDEN, Alvin

Repairs and Refinishing of Gunstocks Was to be one of the *Firearm Design and Assembly* series of booklets. Linden passed away in 1946 before the manuscript was completed.

LINDEN, Alvin

The Design of Gunstocks
and
Gunstock Design and Stockmaking

Both titles refer to the same work, which was intended to be a hardbound volume containing material presented in the first three *Firearm Design and Assembly* series booklets. For some unknown reason, presumably Alvin Linden's death, Samworth elected not to produce a hardcover version. Stackpole Publishing, however, did produce a hardbound volume containing all three published Linden *Booklets*; see Notes for title #46.

In ads, Samworth described this stillborn title:
". . . is a forthcoming textbook which is now being written. It will be the final word in the field of gunstocking, finishing, repair and design. Will include most of the text and illustrations appearing in this present series of booklets plus the addition of much new material. If interested, send in your name for our files, so that an announcement can be mailed to you when this book is ready."

TRADE FROM THE MONONGAHELA

THE RIFLE FROLIC

SAMWORTH PRINTS ON FIREARMS

The dust jacket artwork on Samworth books was often colorful and of particularly high quality. So many comments were received that the publisher began a series of prints based on dust jacket illustrations. This series was envisioned to include most of the detailed works by Gayle Hoskins and E. Stanley (Ned) Smith, both former understudies of Howard Pyle, premier apostle of the "Brandywine" school of artwork. Characteristics of this type of painting include less-than-sharply defined linework and wide use of multicolored tones and hues to emphasize detail. It is interesting to note that this technique was looked down on as being inferior to other contemporary styles of the 1940s and 1950s.

The concept of offering prints was an ostensibly excellent idea, but Samworth halted the offerings after the publication of only two prints. The sideline venture did not prove out to have the popular demand that was believed to exist.

The series is described in Samworth's own words:

SAMWORTH PRINTS

"In the face of continued public demand it is only natural that the pioneer publisher of books on firearms in this country should blaze the trail in making available authentic prints on firearms.

"To launch this series of *Samworth Prints on Firearms* we offer two subjects that will warm the hearts of riflemen in particular and appeal to all sportsmen in general. The prints are reproductions in full color of masterful, authentic historical paintings by Gayle Hoskins, widely known artist. Each measures 17" x 25" plus wide margins and are priced at $5.00 each delivered suitable for framing.

"*Trade From the Monongahela* depicts a bartering scene at an early Colonial riflemaker's establishment in the vicinity of Lancaster, Pennsylvania, about

1750. The bundles of deer hides will be swapped for new rifles, ammunition and components. In shops such as this (it may or may not be that of the re-knowned Phil LeFevre) the American arms industry was born. Here the famous Pennsylvania--in more recent years miscalled Kentucky--rifle was made and developed. This scene is used on the jacket and frontispiece of *The Gunsmith's Manual,* a Samworth reprint first published by Stelle & Harrison in 1882.

The Rifle Frolic, 1775, shows Captain Nagle's Company of Pennsylvania Riflemen holding a shooting match at a crossroads tavern in Berks County. A month or so later, "Nagle's Dutchmen" as they were generally called, were on the march to join up with General Washington's army at Cambridge. There they made history and established rifle scores of which every American is intensely and justly proud. In depicting this rough and ribald lot, all of whom came well within the classification of "Riflemen," the artist deftly and uncannily caught the spirit of the people of that day. This scene will be used on the jacket and as the frontispiece of Ray Smith's forthcoming *Pennsylvania Rifles and Riflemakers.*

About the Artist: Gayle Hoskins, a native of Wilmington, Delaware, was an outstanding student of the late Howard Pyle. He has an especial aptitude for historical scenes and outdoor subjects involving firearms and gives a most satisfying degree of interest, study and execution to even the most critical demands for accurate weapons-and-period correctness. Many gun book buyers have been insistent that certain of Hoskins' shooting scenes be reproduced suitable for framing. This is being done in the *Samworth Prints on Firearms* Series."

Samworth's Prints were lithographed by Beck Engraving of Philadelphia, known widely for their quality work. The late Ray Riling commissioned reprints of both Hoskins prints, but these are approximately 2/3 the size of the original prints. Therefore, collectors should note the illustration size of the original Samworth Prints given above. They are quite scarce.

GLOSSARY

a.e.g. All edges gilt; top, foredge and bottom covered with gold foil or an imitation.

association copy A copy of a book having links to the author, illustrator or another notable individual.

boards The outermost parts of a book, the cover. Boards are either leather, cloth or paper covered. When paper covered, the book is described as being in boards.

book paper An uncoated grade of paper specially designed and produced for books with principally text and little or no illustrations. Some book papers reproduce photographs and artwork with nearly the same fidelity as coated papers.

bookplate A label specially printed to show ownership of a book and normally fixed on the front paste-down.

buckram A coarse weave linen or cotton cloth that may be filled with pigment to varying degrees; used as a hard wearing binding cloth.

calendering The process of imparting gloss to paper by passing the paper web through a stack of rollers, called a calender.

called for A bookseller's term meaning that some item is necessary for the completeness of a book.

cancel A substitute leaf bound in after the sheets of a book have been printed and folded.

cancelland An unaccepable leaf which was removed and replaced by a cancel.

cap The top or bottom of the spine.

case, casing The method of fixing the covers to a book without stitching the lacing cords to the boards. Cased books have the covers attached by gluing reinforcing strips to the endpapers. Modern usage applies this term to all cloth and paper hardcovered books.

coated paper That paper to which a "coating" mixture of pigments and binders is applied. Can be purchased

in finishes including gloss, dull, and matte.
colophon A list of details giving printer, paper stock, typeface, font, number of copies, etc. Usually found at the rear.
cut A book with edges cut has the rough edges of the paper removed and the leaves separated.
deckle The ends of the mold employed by paper-makers who make paper by hand. In modern paper-making, the sides of the forming section of a paper machine.
deckle edge The rough, almost torn edge on paper that is formed near a deckle. It is almost always trimmed off on the paper machinc, but books containing handmade paper almost always display these rough edges.
dull Coated paper that is only slightly glossy, like an eggshell.
dust jacket The paper jacket put around most modern hardcover books for protection and advertising purposes. Also called dust wrapper.
endpapers The blank leaves at the front and back of a book. The endpapers glued to the covers are paste-downs; those immediately adjoining are the free end-papers. All other blank leaves are flyleaves.
errata A loose sheet listing textual errors in a book. A list of errors (pl. errata).
ex-library Strictly, a book marked as having been in a lending library.
filler A colored pigment applied to a cloth substrate manufactured for binding cloth.
folio The number of a page in a book. Also a single-folded sheet. See format.
format The number of times a whole sheet has been folded to provide the leaves of a single gathering for a book. Folios are folded once, quartos twice, octavos thrice, and sextodecimos four times, etc.
foreword Introductory remarks about a book and its author. Not "forward".
foxing Brown spots and blotches caused by a photo-chemical reaction of acid and lignins remaining in the paper.
front matter All pages before the main text.

frontispiece An illustration facing the title page. Seldom spelled out; abbreviation "frontis." is usually seen.
galley The proof typeset sheets prior to printing; sometimes bound and sent to reviewers.
gathering A section of a book formed when the printed sheet is folded the requisite number of times. Also called a signature.
gilt Gold foil, genuine or imitation, used in lettering.
gloss finish Coated paper with a very high "shine" or gloss. Also called "super" finish.
half-binding A binding where leather covers the spine of the book and continues on the front and back covers slightly. Small triangular patches of leather are applied to the corners, which differentiate this binding style from quarter-binding.
half-title The leaf in front of the title page carrying only the book's title.
halftone A black and white photograph prepared for the printing process by being photographed through a fine-line screen. Typical screens are 100 to 300 lines per inch. The screen breaks up continous tones into patterns of tiny dots which are more reproducible.
headband A protective band sewn or glued inside the top and bottom of a spine.
impression The complete number of books printed at one time. The first edition as spoken of by collectors is the first impression.
issue The total number of copies published at one time. This may be only part of an impression, so there can be multiple issues of an impression.
limited edition An edition with a limited number of copies. If a limitation goes above one thousand it loses significance.
matte finish Coated paper with a true flat appearance with no gloss.
monograph A book which deals with one subject or a part thereof.
morocco A grained leather derived from goatskin and often dyed red.
octavo A thrice-folded sheet. See format.

offset lithography An image to be printed is transferred to a rubber blanket and then "offset" on the paper.
perfect binding A glued-on stiff paper cover; commonly called softcover binding. Referred to in the used book trade as wrappers, abb. "wraps."
pigskin A plain but hardwearing covering.
point A vertical unit of measurement used in typesetting. One point equals 1/72 of an inch.
points The features such as misprints, corrections, advertisements, dates, colors, and other physical characteristics used as guides to distinguish one issue or edition from another.
preface The introductory remarks by the author in the front matter, containing the reason for the book and its scope.
presentation copy A copy of a book presented by the author or illustrator and carrying an inscription, not merely signed.
provenance The history of a book's ownership, expressed often by previous owner's marks and bookplates, or references in sale catalogs. Many collectors mistakenly desire to collect only pristine copies, often turning down books with significant provenance.
quarter-binding A style where the leather covers only the spine of a book.
quarto A twice-folded sheet. See format.
raised bands The decorative ridges on the spine of leather-bound books formed by the binding cords not being recessed in the spine.
recto The right-hand page of an open book.
running heads The book or chapter title found at the top of each page.
saddle stitched A binding method where wire staples pass through the spine fold and fasten the cover and pages together. Most magazines and books under 64 pages are saddle stitched.
screen See halftone.
serif The "tails" on a character that make a typeface easier to read.

sic "Intentionally so written." Used after a printed word or passage to indicate that it exactly appears as the original.

signature A letter or mark printed on lower margins of the first (and sometimes subsequent) page of a gathering. Used as a synonym for gathering.

signed copy Usually meaning a copy of book signed by the author or illustrator and not anyone else, no matter how notable.

size A starch or other polymer-based solution applied to paper or cloth to stiffen and waterproof the carrier material.

Smyth sewn Signatures that are sewn together with thread prior to attachment of the cover.

spine The part of the book connecting the front with the back.

stock Paper a book is printed on. "Cover stock" is stiff and heavy, "text stock" is thinner and more flexible.

sunning Bleaching and fading of dust jacket or cover colors by sunlight.

title piece A thin paper or leather label bearing the title and attached to the spine or front cover.

uncut Books which have all edges smooth are said to be trimmed. A book is uncut if all the edges of its leaves are in the rough state.

unopened A book is unopened if the leaves have not been slit apart after folding and before binding.

vellum The smooth inner side of calfskin, nearly white in raw form and usually dyed. When cared for it will outlast leather as a binding cover.

verso The left-hand page of an open book.

vignette A small illustration or decoration used at the beginning or end of a chapter.

woodcut An illustration printed from a block of wood on which a design or image has been cut, characterized by very fine, closely-spaced parallel lines.

working title A preliminary title used to refer to a book during preparation for publication.

wrappers Paper covers as in soft cover paperbacks. Usually abbreviated to wraps.

TRIM SIZES OF BOOK FORMATS

dimensions in inches

Medium Duodecimo	(12mo)	7 1/2 X 4 1/2
Crown Octavo	(8mo)	7 1/2 X 5
Post Octavo	(8mo)	7 1/2 X 5 1/2
Demy Octavo	(8mo)	8 X 5 1/2
Medium Octavo	(8mo)	9 1/2 X 6
Royal Octavo	(8mo)	10 X 6 1/2
Imperial Octavo	(8mo)	11 1/2 X 8 1/4

Octavos are found in such a broad range of sizes because each finished trim size starts with a single sheet of paper of differing size, but all are folded three times to yield 16 pages, eight per side (eight-up).

NOTES

NOTES